Membrane Receptors and Enzymes as Targets of Insecticidal Action

Membrane Receptors and Enzymes as Targets of Insecticidal Action

Edited by
J. Marshall Clark
University of Massachusetts
Amherst, Massachusetts

and
Fumio Matsumura
Michigan State University
East Lansing, Michigan

PLENUM PRESS · NEW YORK AND LONDON

Library of Congress Cataloging in Publication Data

Membrane receptors and enzymes as targets of insecticidal action.

"Proceedings of a satellite symposium of the seventeenth International Congress of Entomology, held August 20-26, 1984, in Hamburg, Federal Republic of Germany" —T.p. verso.
    Bibliography: p.
    Includes index.
    1. Insecticides—Physiological effect—Congresses. 2. Nervous system—Insects—Congresses. 3. Insects—Physiology—Congresses. 4. Insect pests—Control—Congresses. 5. Cell receptors—Congresses. 6. Cell membranes—Congresses. 7. Enzymes—Congresses. I. Clark, J. Marshall (John Marshall), 1949-   . II. Matsumura, Fumio. III. International Congress of Entomology (17th: 1984: Hamburg, Germany)
SB951.5.M45   1986   632′.951   86-4908

Proceedings of a Satellite Symposium of the Seventeenth International Congress of Entomology, held August 20-26, 1984, in Hamburg, Federal Republic of Germany

ISBN-13: 978-1-4684-5115-3   e-ISBN-13: 978-1-4684-5113-9
DOI: 10.1007/978-1-4684-5113-9

© 1986 Plenum Press, New York
Softcover reprint of the hardcover 1st edition 1986

A Division of Plenum Publishing Corporation
233 Spring Street, New York, N.Y. 10013

PREFACE

One of the fundamental concepts of toxicology is that chemicals
act at selective receptors and that such interactions result in phar-
macologic responses which, depending on dose, may or may not result in
toxicity. For us to understand how insecticides produce their toxic
effects, we must first understand their molecular interactions with
their target receptors. With this in mind, we organized a symposium
which was given in conjunction with the XVII International Congress of
Entomology in Hamburg on August 21, 1984. The goal of this symposium
was to bring together researchers with a wide range of expertise who
shared a common interest in the action of insecticides on the insect
nervous system. It was decided to restrict the scope of the symposium
so that selected topics could be discussed in greater depth. The
volume which resulted from this symposium, "Membranes Receptors and
Enzymes as Targets of Insecticidal Action", details a number of biochemical modes of action of insecticides on the insect nervous system.

The volume is divided into two sections; the first dealing with
the action of insecticides on the GABA-chloride channel complex. This
section evolves from a discussion of the symptoms of cyclodiene
toxicity presented by Dr. D.E. Woolley, to the structure-activity
relationships and pharmacology of the channel complex and is concluded
with the extremely interesting work of Dr. C.C. Wang on the action(s)
of avermectin at this receptor.

The second section deals with the action of insecticides on the
neurochemistry and neurophysiology of the acetylcholine receptor, the
glutamate receptor and neurosecretion. Also modulatory events such as
cyclic AMP stimulated and calcium stimulated phosphorylation-
dephosphorylation of proteins in insect nerve membranes are discussed
in terms of sites of action of certain insecticides.

We have attempted in our editorial process to make this collec-
tion of papers as consistant as possible and to organize it with a
continuum of thought. Any errors or omissions are therefore the
collective fault of the editors. We are grateful to the contributors

who really did the bulk of the work and to our typist Ms. Priscilla
Coe.   We also wish to thank the organizing committee of the XVII
International Congress of Entomology for allowing us to publish this
volume and in particular Dr. W. Knauf who served as section organizer.

                               J. Marshall Clark and Fumio Matsumura

AIMS AND SCOPE

The enormous growth which has taken place in our understanding of the insect nervous system due to neurochemical and biochemical approaches has ushered in a new era of insecticide toxicology. As new methods have become available for the study of nerve membranes, it is less appropriate to base our assumptions concerning insecticidal modes of action solely on classical electrophysiological studies.

The aim of this book is to bring together researchers with a vast array of expertise who all are involved in the common pursuit of how insecticides work on the insect nervous system at the molecular level. In this way, each chapter represents the unique approach taken by individual researchers to study this common subject.

The scope of this book encompasses the modes of action of insecticides that act directly on the insect nervous system. The action of such insecticides are examined in detail on the GABA-chloride channel complex, the acetylcholine receptor and the glutamate receptor. Additionally, the action of insecticides on various secondary messengers, such as cyclic AMP and calcium, are covered in the consideration of neurosecretory processes, calcium regulatory processes and membrane phosphoryation-dephosphoryation events.

CONTENTS

EFFECTS AND PROPOSED MECHANISMS OF ACTION OF LINDANE IN MAMMALS:

UNSOLVED PROBLEMS

Dorothy E. Woolley and Louis Zimmer

Departments of Animal Physiology and Environmental
Toxicology
University of California
Davis, California 95616

INTRODUCTION

The several objectives of this brief review are: 1. to
describe some of the therapeutic uses of lindane in man and domestic
animals and the potential for toxicity that results from this use;
2. to describe the major signs of toxicity produced by lindane in
man and other mammals; 3. to review the proposed mechanisms of
action of lindane at the cellular level; 4. to consider whether or
not the proposed mechanisms of action can, in fact, explain the signs
of poisoning produced by lindane in mammals; and 5. to point out the
many similarities in effects and proposed mechanisms of action
between lindane and some cyclodiene insecticides, such as dieldrin,
even though the chemical structures of lindane and the cyclodienes
appear dissimilar.

REVIEW OF RELATED WORK

Effects of Lindane in Man and Other Mammals

Lindane, the gamma isomer of hexachlorocyclohexane, continues to
be used topically in both human and veterinary medicine to treat
ectoparasites such as mites and lice (Soloman et al., 1977; Ulmann,
1972). In 1948, a cream containing 1% lindane was used successfully
to treat scabies in man (Wooldridge, 1948; Halpern et al., 1950) and
this treatment continues to be widely used today (Solomon et al.,
1977). Properly used, it is considered safe since the dermal
absorption of lindane is low -- with some exceptions. Lindane
penetrates the skin of infants far more readily than the skin of

1

adults; a recent report described an infant found dead in its crib after treatment for scabies with a lindane solution (Davies et al., 1983). Also, some regions of the skin, such as the scrotum, offer little barrier to penetration. Furthermore, lindane would be expected to penetrate excoriated skin, such as may be found in association with the ectoparasite infections for which it is used (Solomon et al., 1977). Thus, there is a very real potential for poisoning by lindane, even in association with its recommended medical usage. In addition, lindane is also used as a general insecticide, particularly in countries outside of the United States (Ulmann, 1972).

The signs of poisoning by lindane have been described for both man and a number of other mammals (Winteringham and Barnes, 1955; Herbst and Bodenstein, 1972; Solomon et al., 1977; Joy, 1982). Since its first use soon after World War II, numerous reports of major toxicity and death in man, associated with accidental or deliberate exposure, have appeared. In 1951, a lindane emulsion was given orally as a vermicide to a number of patients (Graeve and Herrnring, 1951). One exhibited convulsions, while others, who received about 0.5 mg/kg body weight for 3 days, complained only of nausea. In the most severe incident (Khare et al., 1977), epidemic poisoning occurred in India when seed grains treated with lindane were used as food grains instead and consumed. The most prominent sign of poisoning was the sudden appearance of various types of seizures, i.e., myoclonus, petit mal and grand mal. Other signs of poisoning included intention tremors, impairment of memory, irritability and aggression. The seizures were successfully controlled with primidone -- a congener of phenobarbital -- and phenytoin.

In the rat and other mammals, seizures are also the principal signs of poisoning by lindane (reviewed by Joy, 1982; Joy et al., 1982; Woolley et al., 1984, 1985). In addition, lindane in subconvulsant doses has proconvulsant effects, i.e., lindane enhances susceptibility to seizures to such an extent that excitation from a source not normally convulsant will now precipitate convulsions (Hulth et al., 1976; Joy et al., 1982). The proconvulsant effects of lindane have been demonstrated by its effectiveness in facilitating kindling (Joy et al., 1982, 1983; Stark et al., 1983). Kindling is the process whereby repetition of an initially subconvulsant electrical stimulation to the brain, especially to certain areas of the limbic system, eventually produces seizures. Kindling has been established as a useful model to evaluate the convulsant and anticonvulsant potential of various chemical agents (Racine, 1978). Like lindane, dieldrin has also been shown to facilitate kindling (Joy et al., 1980).

An extensive review of the toxicity of lindane in a number of species concluded that domestic animals were more sensitive to the toxic effects of both dermal and oral administration of lindane than

were rats (Herbst and Bodenstein, 1972). A single topical
application of 1% lindane to weanling rabbits, at a dose reportedly
used clinically in infants, produced convulsions and severe anorexia,
whereas the same dose per unit body weight to young adult rabbits
produced only some anorexia and possibly mild excitement (Hanig et
al., 1976).

In summary, the effects of lindane in man and other mammals
include not only the well known excitatory and convulsant effects but
also include additional effects such as anorexia and impairment of
memory. The purpose of the studies to be described is to compare the
convulsant effects of lindane with other effects, i.e., body
temperature, food intake, body weight, neurophysiological responses,
to see if some of these effects can be explained in terms of the
proposed mechanisms of action.

Proposed Mechanisms of Action

The action of lindane was first studied in insects and shown to
be on the ganglia rather than on the isolated nerve cord (Yamasaki
and Ishii, 1954; Narahashi, 1971). The excitatory action of lindane
on the cholinergic synapse in the giant nerve cord of the cockroach
was thought to be due to a presynaptic action resulting in an excess
release of acetylcholine (ACh) because the effects were blocked by
nereistoxin, which allosterically blocks the response of the receptor
to ACh (Uchida et al., 1975). Lindane increased the spontaneous
release of neurotransmitter at the isolated neuromuscular junction of
the frog, as demonstrated by the increased occurrence of miniature
endplate potentials (MEPPs) after exposure of the preparation to
lindane (Publicover and Duncan, 1979). The effect was believed to be
due primarily to an increase in Ca^{++} permeability and increased entry
of Ca^{++}, since the major part of the effect was blocked by removal of
extracellular Ca^{++}. A small increase in MEPP frequency after
exposure to lindane persisted in the absence of extracellular Ca^{++}
and was attributed to a lindane-induced release of Ca^{++} from
intracellular stores. Dieldrin also increased neurotransmitter
release (Wang et al., 1971; Shankland and Schroeder, 1973; Joy,
1982). Recently, Joy (1982) proposed that the basic action of both
lindane and dieldrin is to intensify synaptic activity, regardless of
whether the synapse is excitatory or inhibitory.

An action of lindane in increasing Ca^{++} entry has also been
suggested by other studies. Narbonne and Lievremont (1983)
demonstrated that the uptake of Ca^{++} by synaptosomes was increased
very rapidly after introduction of lindane into the medium. This was
true whether the synaptosomes were in a medium with normal (5 mM) or
depolarizing (71 mM) levels of K^+. In view of the well established
relation between levels of Ca^{++} within presynaptic nerve terminals
and neurotransmitter release (Katz and Miledi, 1970), these
observations provide further support for the conclusion that the

lindane-induced increase in MEPP frequency at the neuromuscular junction of the frog was indeed due to increased Ca^{++} entry into presynaptic terminals (Publicover and Duncan, 1979). Increased intracellular Ca^{++} ions are also known to be necessary for other forms of exocytosis as well. Lindane-induced release of the neurosecretory products of the insect corpus cardiacum, accompanied by many exocytotic omega figures and ultrastructural damage in mitochondria, has also been ascribed to excessive entry of Ca^{++} (Normann and Samaranayaka-Ramasamy, 1977). However, the relation between lindane and Ca^{++} entry or release from intracellular stores is apparently not simple. Recently, Lievremont et al. (1984) have provided evidence that lindane inhibits Ca^{++} entry into skeletal muscle.

A new approach to the effects of lindane and other organo-chlorine insecticides was provided when Matsumura and his colleagues pointed out the similarity between the action of lindane and some cyclodiene insecticides and the action of picrotoxinin (Ghiasuddin and Matsumura, 1982; Kadous et al., 1983; Matsumura and Ghiasuddin, 1983; Matsumura and Tanaka, 1984). Picrotoxinin inhibits GABAergic transmission by binding to a receptor site located on the GABA-activated chloride channel in the nerve membrane (Olsen et al., 1984). Lindane competed with picrotoxinin for binding to the receptor site in synaptosomal fractions from rat brain (Matsumura and Ghiasuddin, 1983). Lawrence and Casida (1984) used t-butylbicyclophosphorothionate (TBPS) as a ligand for the picro-toxinin receptor and found that 3 major classes of organochlorine insecticides, i.e., lindane, cyclodienes and toxaphene, exhibited potent competitive inhibition of TBPS for the binding site. Further-more, the binding affinity for this site correlated well with in vivo toxicity. Similarly, work in the laboratory of the Eldefrawi's (Abalis et al., 1985; Eldefrawi et al., 1985) has confirmed that lindane and several cyclodienes compete with TBPS for binding to the picrotoxinin receptor site in membranes from rat brain in vitro and that binding affinities in vitro correlate well with toxicities in vivo. The observation that lindane did not displace either GABA or flunitrazepam from binding to membranes from rat brain (Cattabeni et al., 1983) does not contradict the reports that lindane does compete with agonists at the picrotoxinin site, since GABA, benzodiazepines, barbiturates and picrotoxinin each bind at different, although closely related, sites on the GABA-receptor/chloride ionophore complex (Olsen, 1982; Trifiletti et al., 1984; Seifert and Casida, 1985). These findings are very compelling because GABA is the major inhibitory neurotransmitter in the mammalian brain and interference with GABAergic inhibition could readily account for the convulsant action of lindane.

EXPERIMENTAL

Specific Objectives

The first objective was to determine the time course of several simple toxicological and physiological effects, i.e., seizures, hypothermia, weight loss and anorexia, produced by a single dose of lindane in the rat. Diazepam and phenobarbital, which enhance GABAergic transmission (Olsen et al., 1984), were administered to lindane-treated rats, in order to evaluate their efficacy in counteracting some of the effects of lindane. This was done because of reports (see preceding section) that lindane inhibits GABAergic transmission. The second objective was to determine the effects of lindane on certain neurophysiological parameters of specific brain systems, in order to gain additional information regarding brain areas affected by lindane and the relation of these to the signs of poisoning. To do this, the effects of lindane on averaged evoked potentials were studied in rats with electrodes chronically implanted in specific areas of the limbic system, so that animals could be studied in the unanesthetized, awake, freely-behaving state. Limbic brain areas were studied because of the known relation between seizures and the limbic system and because of the established convulsant effects of lindane (Woolley et al., 1984). In addition, paired pulse stimulation was used to evaluate the integrity of GABAergic recurrent inhibition.

Effects of Lindane on Seizures, Food Intake and Colonic Temperatures

Methods. Adult female, Sprague-Dawley rats (Simonsen Laboratories: Gilroy, Calif.) were used. After an overnight fast, lindane (Sigma; St. Louis, Missouri) dissolved in corn oil was administered per os (po) by intubation in a volume of 1 ml/kg body weight. Preliminary range-finding studies demonstrated that a single dose of 30 or 40 mg/kg body weight of lindane produced measurable but non-lethal effects. Of approximately 100 rats dosed with the higher dose, only 1 death occurred. Colonic temperatures were measured with a rectal thermister probe (Yellow Springs Instrument Tele-Thermometer Model 42SC; Yellow Springs, Ohio). When food intake was measured, rats were fed ground rodent chow (Purina 2001) in food cups fixed within an outer cup to catch spillage. Otherwise, the chow was in pelleted form and available ad libitum. Water was available at all times. Diazepam was dissolved in dimethylsulfoxide (DMSO) and injected subcutaneously (sc) in a dose of 5 mg/kg. Sodium phenobarbital was dissolved in distilled water and injected sc in a dose of 30 mg/kg. Both drugs were injected in a volume of 1 ml/kg. Animals not injected with a drug received an injection of the vehicle instead at the same time periods. There were 6-9 rats per group.

Results. After a single dose of lindane (40 mg/kg) in oil
administered per os, plus DMSO sc, seizure activity was usually first
observed between 15-30 min later, was maximal at about 45-90 min, and
had almost disappeared by 2.5 hr (Fig. 1). However, seizures were
still occasionally observed 24 hr later. The first sign of seizure
activity was myoclonus, consisting of a single, whole-body jerk,
which progressed to more frequent jerks. As seizure activity
intensified, clonic movements of the forelimbs and head appeared, as
well as rearing sometimes with loss of equilibrium. A more severe
clonic seizure was characterized by an initial tonic extension of the
head and loss of equilibrium, followed by clonic seizure activity.
Tonic seizures with tonic flexion followed by extension of the
forelimbs occurred less frequently (Fig. 1). Administration of
either diazepam or phenobarbital simultaneously with administration
of lindane prevented the appearance of any seizure activity.

Lindane produced pronounced anorexia (Figs. 2 and 4) and
hypothermia (Fig. 3). Following return of food, after the
pretreatment overnight fast and 4 hr after administration of lindane
and/or drugs, control and drug-treated animals ate, whereas rats
receiving only lindane did not. Food intake was severely depressed
for 3 days after administration of the single dose of lindane (Figs.
2 and 4) and body weights were significantly depressed for 4 days
(Fig. 2). Administration of diazepam at 12 hr intervals for 3 days,
i.e., for 6 injections, blocked the effects of lindane on body weight
and food intake (Fig. 2). Similarly, administration of phenobarbital

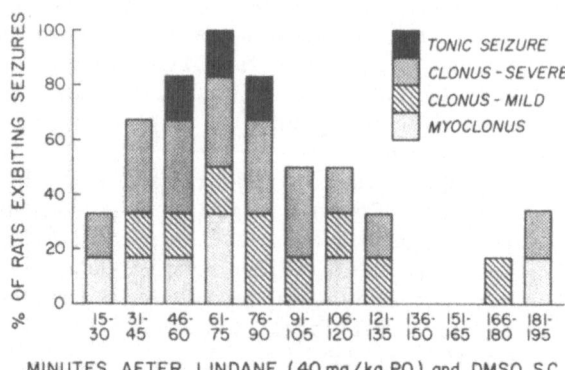

Fig. 1. Time course of overt, spontaneous seizure activity following
        a single dose of lindane dissolved in oil. DMSO was
        injected subcutaneously as a vehicle control for the drug-
        treated animals, shown in Figs. 2 and 3. Only the most
        severe response for each rat in a given 15 minute block of
        time was plotted. See text for description of categories of
        seizure activity.

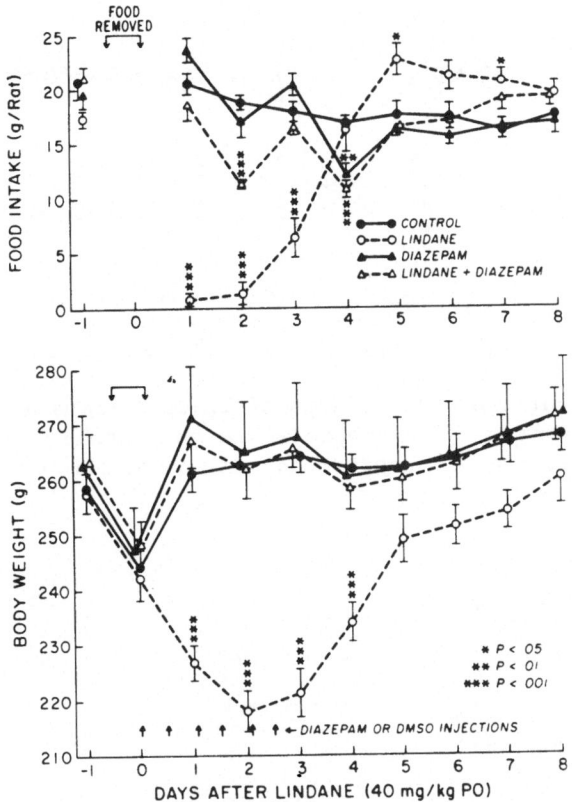

Fig. 2. Time course of the effects of lindane alone or with diazepam injections or of diazepam alone on body weights and food intake. Diazepam (5 mg/kg) was injected subcutaneously in DMSO twice daily for 3 days. Asterisks indicate that values are significantly different from control values. Reproduced with permission from Woolley et al. (1985).

blocked the anorexic effects of lindane (Fig. 4) and the effects on body weight (not shown).

A single administration of lindane produced a rapid and marked fall in colonic temperatures; colonic temperatures dropped more than 2°C in 45 min, were even lower at 2 hr, and were still significantly depressed 2 days later (Fig. 3). The hypothermia produced by lindane was prevented by administration of diazepam, even though diazepam alone produced a slight hypothermia (Fig. 3). Phenobarbital also effectively blocked the hypothermic effects of lindane (not shown).

Summary. Although the major action of lindane is generally considered to be that of a convulsant, its effects in producing

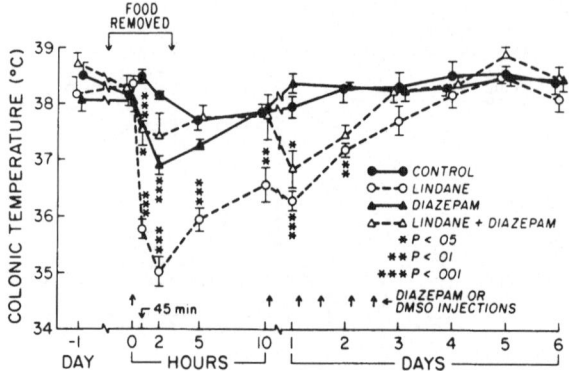

Fig. 3.   Time course of the effects of lindane alone or with diazepam
          injections or of diazepam alone on colonic temperatures.
          Also see Fig. 2. Reproduced with permission from Woolley et
          al. (1985).

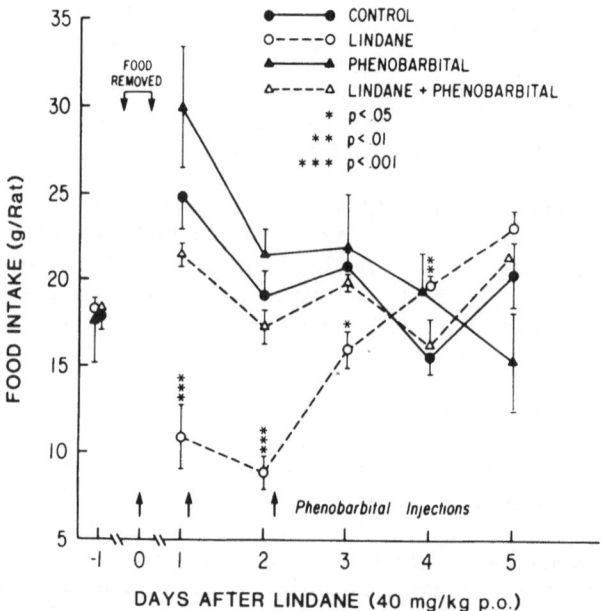

Fig. 4.   Time course of the effects of lindane alone or with
          phenobarbital injections or of phenobarbital alone on food
          intake. Sodium phenobarbital was dissolved in saline and
          injected (30 mg/kg) subcutaneously once daily for 3 days.
          Asterisks indicate that values are significantly different
          from control values. From Woolley and Dodge, unpublished
          observations.

hypothermia and anorexia were quite severe and lasted far longer than did its effects in producing overt seizures. All these effects of lindane, i.e., seizures, hypothermia, anorexia, were blocked by diazepam and phenobarbital, indicating that they can be prevented by enhancement of GABAergic transmission.

Effects of Lindane on Limbic Evoked Potentials

Methods. To study evoked potentials, bipolar, side-by-side electrodes with approximately 1 mm vertical tip separation (Woolley, 1977) were positioned in limbic brain areas (Fig. 5) in anesthetized rats, according to general procedures described previously (Hasan et al., 1984). Electrodes were connected to the pins of a subminiature connector so that, following recovery from surgery, the animals could be "plugged in" to stimulating and recording equipment for study in the freely behaving state. An electrode was used for either stimulation and/or recording as required. Electrophysiological responses were evoked in the dentate gyrus (DG) of the hippocampal formation either by stimulation of the olfactory or prepyriform cortex (PPC) or by stimulation of the dorsal perforant path (PP). Responses in the CA3 subfield of the hippocampus were evoked by stimulation of the DG. Ten to 30 responses were averaged with a Data General Nova 1200 computer to produce an averaged evoked potential (AEP). Prior to recording AEPs, animals were allowed 10 days for recovery from surgery. Then AEPs were recorded 3 times a week for 2 weeks, then daily until AEPs had stabilized. Animals were then fasted overnight and intubated with the oil vehicle to study the effects of this treatment alone on AEPs, recorded according to the same protocol used later when lindane was administered. Because lindane in a dose of 40 mg/kg was shown to produce hypothermia and because hypothermia itself can affect evoked potentials, a dose of 30 mg/kg of lindane was used to study effects on evoked potentials. This lower dose in oil administered po was shown not to produce hypothermia (Zimmer and Woolley, unpublished observations). Also, severity of seizures was far less at this dose and 25 percent of the animals did not show overt seizures.

Background for Interpreting Responses Evoked in the Denate Gyrus (DG) and CA3. Fig. 6 is presented as an aid in interpreting the evoked responses elicited in the dentate gyrus. It illustrates a granule cell, the major neuron type in the dentate gyrus (also see Fig. 5), and the terminations of the perforant path, which is the major input to the dentate gyrus. The perforant path is excitatory to the dentate gyrus (Andersen et al., 1966; Lomo, 1971) and the neurotransmitter involved is probably glutamate (Storm-Mathisen, 1977). The pathway between the prepyriform cortex and dentate gyrus contains a synapse in the lateral portion of the entorhinal cortex (Fig. 5). The path continues on to the dentate gyrus from the lateral entorhinal cortex via the perforant path and terminates in the outermost layers of the dendrites, farthest from the cell bodies

of the granule cells (McNaughton and Barnes, 1977; Wilson and
Steward, 1978). By contrast, the fibers from the medial entorhinal
cortex, which also travel in the perforant path, terminate in the

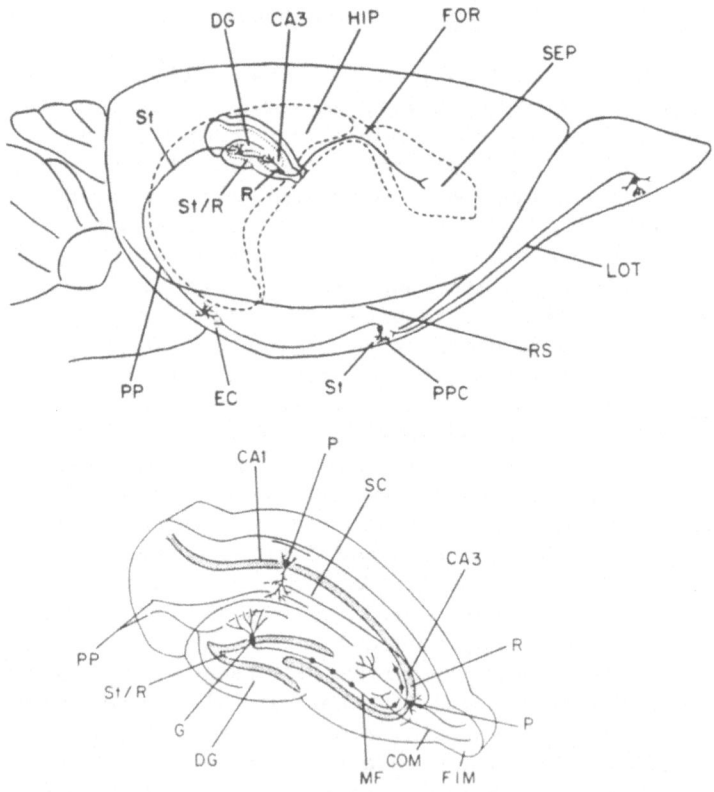

Fig. 5.   Top. Side view of the rat brain to show the relation
          between the olfactory or prepyriform cortex (PPC), entor-
          hinal cortex (EC), perforant path (PP), and hippocampal
          formation (HIP), which includes the dentate gyrus (DG) and
          hippocampus proper.   Bottom. An enlargement of a section
          through the dorsal hippocampal formation to show more detail
          of the neuronal organization. The granule cells (G) are the
          principal neuron type of the DG and pyramidal cells (P) are
          the major neuron type in the CA1 and CA3 subfields of the
          hippocampus proper. Stimulating (St) and recording (R)
          positions used to elicit and record evoked potentials are
          indicated in both views. Axons of the granule cells form
          the mossy fiber (MF) system which activates pyramidal
          neurons in CA3. Other abbreviations are: COM--commissural
          fibers, FIM--fimbria, FOR--fornix, LOT--lateral olfactory
          tract, RS--rhinal sulcus, SC--Schaffer collaterals, SEP--
          septum. Redrawn from Hasan et al. (1984) with permission.

middle of the dendritic tree, much closer to the cell body layer
(Fig. 6). Thus, stimulation of the medial entorhinal cortex has been
shown to be much more effective in causing the granule cells to fire,
i.e., to elicit action potentials along the granule cell axons, than
has stimulation of the lateral entorhinal cortex (McNaughton and
Barnes, 1977; Wilson and Steward, 1978). Similarly, because the
lateral entorhinal cortex provides the input to the dentate gyrus
from the prepyriform cortex, stimulation of the prepyriform cortex is
less likely to cause the granule cells to fire than is stimulation of
either the medial entorhinal cortex or the perforant path directly,
because the latter would include stimulation of fibers from the
medial entorhinal cortex. In addition, direct stimulation of the
perforant path activates the granule cells monosynaptically, whereas
the input to the dentate gyrus from the prepyriform cortex is
disynaptic.

Thus, stimulation of the perforant path, either directly or
indirectly via stimulation of the prepyriform cortex, evokes
excitatory postsynaptic potentials (EPSPs) in the dendrites. The
evoked potentials recorded with macroelectrodes, such as used in the
current study, represent the summation of the EPSPs produced in a
population of neurons and so are called <u>field</u> or <u>populations EPSPs</u>
(Figs. 7, 9 and 11). They are also sometimes called <u>population slow</u>
<u>waves</u>, to contrast them with the very short duration <u>population</u>

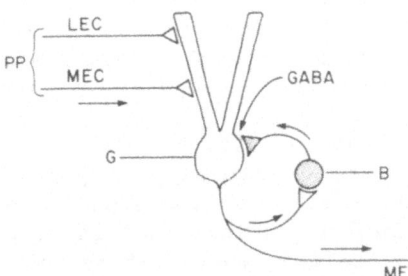

Fig. 6. A schematic view of a typical granule cell (G) in the
         dentate gyrus to show the terminations in the dendrites of
         the inputs from the lateral (LEC) and medial (MEC)
         entorhinal cortex, which reach the granule cells via the
         perforant path (PP). The input to the granule cells from
         the prepyriform cortex (PPC) is carried in fibers from the
         LEC. Axons of the granule cells, which provide an
         excitatory input to CA3 neurons, are called mossy fibers
         (MF). Collaterals of granule cell axons activate GABAergic
         inhibitory interneurons, called basket cells (B), which feed
         back to the granule cells to produce long-lasting inhibition
         of firing of the granule cells.

spikes (PSs), which represent the synchronous firing of action
potentials by a population of neurons (Figs. 7, 9 and 11). When the
population EPSPs are large enough, the granule cells "fire", i.e.,
set off action potentials, as indicated by the appearance of
population spikes. When the granule cells fire, recurrent axon
collaterals activate the GABAergic basket cells, which feed back to
the granule cells (Fig. 6) to produce long-lasting inhibitory
postsynaptic potentials (IPSPs) (Andersen et al., 1966).

By applying two successive, high intensity electrical stimuli
directly to the perforant path, it is possible to determine if
recurrent inhibition in the dentate gyrus is functional. The first
stimulus evokes a high intensity population EPSP in the dendrites,
which in turn causes the neurons to fire. Firing of the neurons is
revealed by the appearance of a population spike during the
population EPSP. Firing of the neuron activates recurrent
collaterals which in turn activate the inhibitory basket cells.
These feed back to the neuron cell bodies, near the site at which the
action potential is normally initiated (Fig. 6), and produce IPSPs
which prevent the cells from firing in response to the second
stimulus. The result is that a population spike will be evident
within the first population EPSP, but not within the second, if
recurrent inhibition is functional. The second population EPSP is
not prevented by recurrent inhibition to the cell bodies because the
EPSPs occur in the dendrites. On the other hand, if inhibition is
reduced or abolished, the population spike will occur during the
second population EPSP, as well as during the first. Because the
input from the prepyriform cortex to the dentate gyrus is disynaptic
and terminates in the outer dendritic layers, stimulation of the
prepyriform cortex will not normally produce a population EPSP of
sufficient intensity to elicit a population spike.

Recurrent inhibition may also be evaluated by means of responses
evoked in the CA3 subfield of the hippocampus. This was done by
using paired pulse stimulation of the granule cells in the dentate
gyrus and recording in CA3, as indicated in Fig. 5. Axons of the
dentate granule cells, called mossy fibers, provide the major
monosynaptic excitatory input to the pyramidal cell neurons of CA3.
When the pyramidal cells fire, recurrent collaterals activate
GABAergic basket cells which then feed back to the pyramidal cells to
provide an inhibitory input, just as do the basket cells in the
dentate gyrus.

Results. In order to study the effects of lindane on
electrophysiological responses elicited in the dentate gyrus and CA3,
paired pulse stimuli were delivered to the prepyriform cortex,
perforant path and dentate gyrus. Examples of the types of responses
evoked in each case, the effects of lindane on these responses, and
the statistical analyses of the effects of lindane are shown in Figs.
7-12.

When the prepyriform cortex was stimulated with two successive electrical stimuli, with varying time intervals between the two stimuli, it was observed that the amplitude of the second slow wave response (R2), recorded in the dentate gyrus, was always higher than the amplitude of the first response (R1), at all intervals ranging from 12 to 256 msec. That is, paired pulse potentiation occurred, so that the population EPSPs were greater after the second stimulus than after the first. The greatest degree of potentiation, i.e., the greatest ratio of the amplitude of R2 to R1, occurred at an interstimulus interval of 40 msec, and so this interval was used to evaluate the effects of lindane. During the pretreatment period, the

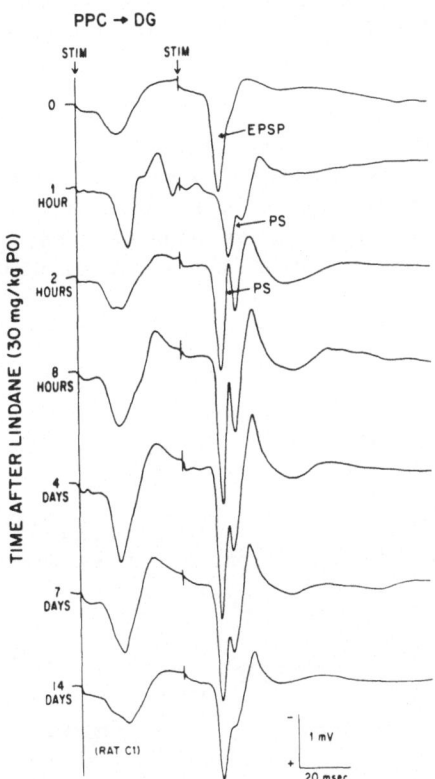

Fig. 7.   The effects of a single dose of lindane on evoked potentials produced in the dentate gyrus (DG) by paired pulse stimulation (STIM) of the prepyriform cortex (PPC). The slow waves represent the summation of excitatory postsynaptic potentials (EPSP) from the population of neurons. After administration of lindane, a population spike (PS) interrupted the population EPSP and indicates that the neurons fired. Some evidence of the PS persists for 14 days. Observations by Zimmer and Woolley. Reproduced with permission from Woolley et al., 1985.

amplitude of R2 was four times greater than that of R1 (Fig. 7) on
the average. One hr after administration of lindane, the amplitude
of R1 was increased nearly 2-fold and by 12 hr it had more than
doubled (Fig. 8). The amplitude remained significantly elevated over
pretreatment levels for six days in most animals, but in some animals
the amplitude was still increased 2 weeks later. The amplitude of
the second slow wave also increased after lindane, although the
percentage increase was not quite as great as that for the first
(Fig. 8). Even though the amplitude of R2 was four times greater
than that of R1 in the control period, it was not sufficient to fire
the granule cells and a population spike was not evident (Fig. 7).
However, by one hr after lindane administration, a population spike
had appeared near the peak of R2, thus indicating that the cells had
fired (Fig. 7). This population spike was sometimes still evident
two weeks later.

When the dorsal perforant path (DPP) was stimulated with two
successive stimuli having an interstimulus interval of 20 msec,
amplitudes of the two population EPSPs elicited in the dentate gyrus
were greater than those elicited by stimulation of the prepyriform
cortex (compare Figs. 7 and 9), as expected (see preceding section).
For this reason, the amplitude of the first population EPSP was

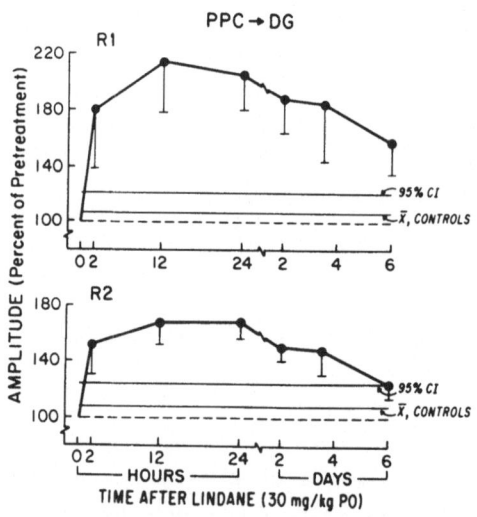

Fig. 8. Changes in the amplitudes of the first and second (R1 and
         R2) population slow waves or population EPSPs, expressed as
         the mean and standard error of the percent of the
         pretreatment amplitudes, after a single dose of lindane.
         Observations by Zimmer and Woolley. Reproduced with
         permission from Woolley et al., 1985.

sufficient to fire the granule cells of the dentate gyrus, as
indicated by a population spike interrupting the initial phase of the
first response (Fig. 9). The population spike was absent during the
second population EPSP, indicating that firing of the granule cells
during the first population EPSP had activated recurrent inhibition
which in turn prevented firing of the granule cells in response to
the second stimulus (Fig. 9). In contrast to the effects on the
responses in the dentate gyrus evoked by stimulation of the
prepyriform cortex, administration of lindane did not dramatically
alter dentate responses evoked by stimulation of the perforant path
(Figs. 9 and 10). Thus, the population spike continued to be present
during the first population EPSP, but was not present during the
second, even after administration of lindane. The amplitude of the

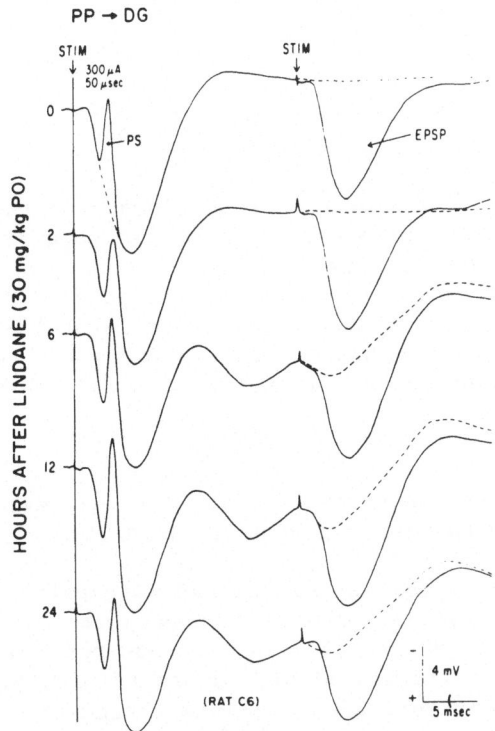

Fig. 9.   An example of the effects of a single dose of lindane on
          evoked potentials produced in the dentate gyrus (DG) by
          paired pulse stimulation of the perforant path (PP). After
          the first stimulus a population spike (PS) interrupts the
          population EPSP but the PS is not evident in the population
          EPSP evoked by the second stimulus. The PS indicates that
          the neurons have fired after the first stimulus but not
          after the second. Observations by Zimmer and Woolley.
          Reproduced with permission from Woolley et al., 1985.

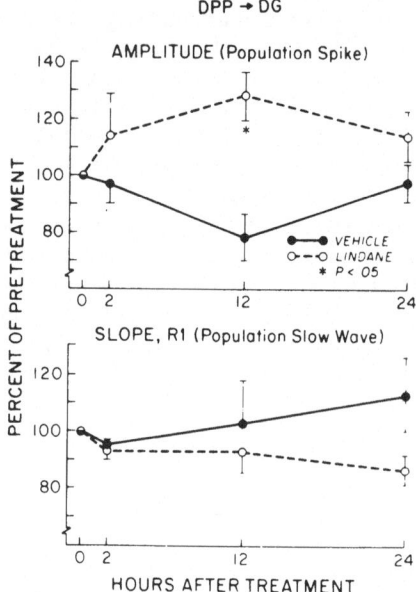

Fig. 10. The effects of a single dose of lindane on the amplitudes of
        the population spike and of the first population EPSP (R1)
        of the potentials· evoked in the dentate gyrus (DG) by
        stimulation of the dorsal perforant path (DPP). Values are
        presented as means and standard errors of the percent of
        pretreatment amplitudes. Observations by Zimmer and
        Woolley. Reproduced with permission from Woolley et al.,
        1985.

population spike increased slightly after lindane, even though the
amplitude of the population EPSP did not (Fig. 10).

     When the dentate gyrus was stimulated with paired pulses having
an interstimulus interval of 20 msec, a population spike interrupted
the first population EPSP recorded in CA3, but not the second. This
was true both in the pretreatment period and after administration of
lindane (Fig. 11). However, after administration of lindane,
additional population spikes appeared during the first population
EPSP (Fig. 11). On the average, lindane slightly increased the
amplitude of the population spike, but not the amplitude of the
population EPSP (Fig. 12).

     Summary. Of the three types of limbic evoked potentials
studied, only the response evoked in the dentate gyrus by stimulation
of the prepyriform cortex was potentiated after administration of
lindane. Also, after lindane, population spikes appeared in this
response whereas there had been none before treatment. Potentiation

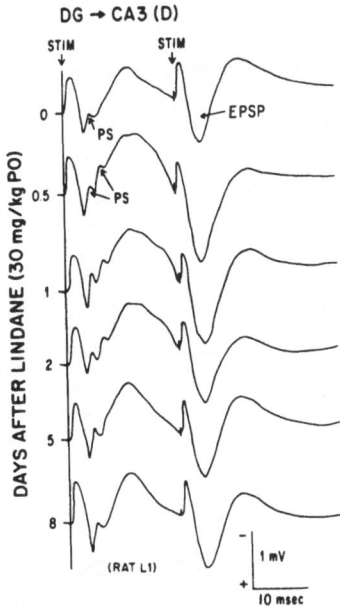

Fig. 11. An example of the effects of a single dose of lindane on
        evoked potentials produced in the CA3 subfield of the
        hippocampus by paired pulse stimulation of the dentate gyrus
        (DG). The first population EPSP is interrupted by a
        population spike (PS) but the second is not. After lindane,
        additional PSs appear during the first population EPSP.
        Unpublished observations by Zimmer and Woolley.

of the response in the dentate gyrus evoked by stimulation of the
prepyriform cortex lasted longer than did effects on body
temperatures and food intake after administration of lindane, even
though a lower dose of lindane was used for the evoked potential
studies. In both the response evoked in the dentate gyrus by
stimulation of the perforant path and the response evoked in CA3 by
stimulation of the dentate gyrus, population spikes were present in
the first population EPSP, but not the second, during paired pulse
stimulation. This was not affected by lindane and indicates that
GABAergic recurrent inhibition remained functional after
administration of lindane.

DISCUSSION

Effects of Lindane on GABAergic Systems

     Comparison of the toxicological effects of a single dose of
either 30 or 40 mg/kg of lindane reveals two interesting and

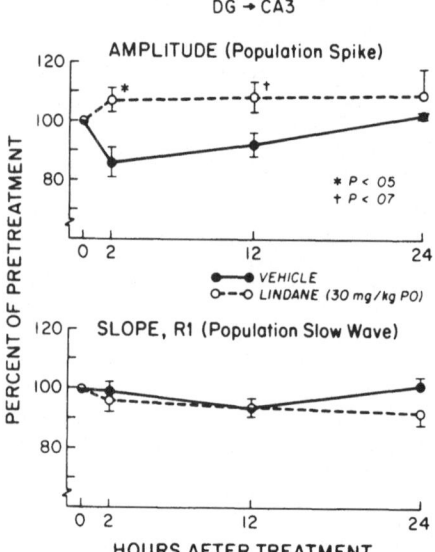

Fig. 12. The effects of lindane on the amplitudes of the population
spike and the first population EPSP (R1, population slow
wave) of the response evoked in CA3 by stimulation of the
DG. Values are means and standard errors of the percent of
pretreatment amplitudes. Unpublished observations by Zimmer
and Woolley.

apparently conflicting findings. First, treatment with either
diazepam or phenobarbital, which enhance GABAergic transmission,
prevented the effects of a single dose of 40 mg/kg on seizures,
hypothermia and anorexia. This would agree with the evidence
reviewed in an earlier section that lindane acts as an agonist on the
picrotoxinin receptor to reduce GABAergic inhibition. On the other
hand, the evoked potential studies revealed that recurrent GABAergic
inhibition in the dentate gyrus and CA3 remained intact after 30
mg/kg. A possible explanation is that only the higher dose of
lindane interfered with GABAergic transmission. However, if this is
so, then lindane-induced seizures can occur at a time when recurrent
GABAergic inhibition in the dentate gyrus is functional. This was
shown by the fact that most of the rats used to study the effects of
lindane on evoked potentials exhibited mild seizure activity at the
same time, i.e., 1 hr after administration of lindane, that recurrent
inhibition in the hippocampal formation was shown to be functional,
as we have reported previously (Woolley and Zimmer, 1984; Zimmer and
Woolley, 1984). Similarly, Joy and Albertson (1984) found that
recurrent inhibition in the dentate gyrus was not reduced by lindane,
even in doses which precipitated convulsions. Therefore, elimination
of recurrent GABAergic inhibition in the dentate gyrus and/or CA3

does not seem to be a prerequisite for mild lindane-induced seizures, although a contribution to more severe seizures remains a possibilty.

Perhaps reduced recurrent inhibition in the hippocampal formation is not a good index of increased susceptibility to seizures. Kindled seizures produced by stimulation of the amygdala were found to be associated with markedly enhanced GABA-mediated recurrent inhibition in the dentate gyrus and with increased benzodiazepine receptor density bilaterally in the hippocampal formation. These changes were believed to be caused by the seizures and to represent an adaptive process to stabilize neuronal excitability to compensate for the increased susceptibility to seizures during kindling (McNamara, 1984). Similarly, Cattabeni et al. (1983) reported that a single administration of lindane in the rat increased turnover of GABA in the cerebellum, cortex and hippocampus, the three brain areas in which it was measured, but only after the onset of seizures. They concluded that once the seizures had started, the GABAergic system was activated, probably as a protective mechanism, as was also suggested for kindling.

Recently, a GABAergic pathway into the hippocampal formation from the basal forebrain was described in the rat (Kohler et al., 1984), as indicated in Fig. 13. The exact function and neuronal termination of this pathway and the type of receptor utilized, i.e., $GABA_A$ or $GABA_B$ receptors, are not yet known. It may have different characteristics than the GABAergic pathway intrinsic to the hippocampal formation which mediates recurrent inhibition. GABAergic recurrent inhibition utilizes the "classical" $GABA_A$ receptor which activates the chloride ionophore and interacts allosterically with benzodiazepines, barbiturates, and picrotoxinin (Olsen, 1982; Olsen et al., 1984). $GABA_B$ receptors have been described only recently, are presynaptic and are activated by calcium ions (Hill and Bowery, 1981). The extrinsic GABAergic path to the hippocampus shown in Fig. 13, by contrast with the intrinsic GABAergic path shown in Fig. 6, may be a logical candidate to use a $GABA_B$ receptor. Therefore, an effect of lindane on this GABAergic pathway remains an interesting possibility (see next section).

Lindane and Long-Term Potentiation of Limbic Evoked Potentials

By far the most dramatic effects of lindane on evoked potentials in the limbic system were the effects on the response evoked in the dentate gyrus by stimulation of the prepyriform cortex, as we have noted previously (Woolley and Zimmer, 1984; Woolley et al., 1984, 1985; Zimmer and Woolley, 1984). The effects of lindane on this evoked response are very similar to the effects of dieldrin (Swanson and Woolley, 1980; Woolley et al., 1984, 1985). The amplitude of both the first and second population EPSPs of the response evoked in the dentate gyrus by paired pulse stimulation of the prepyriform

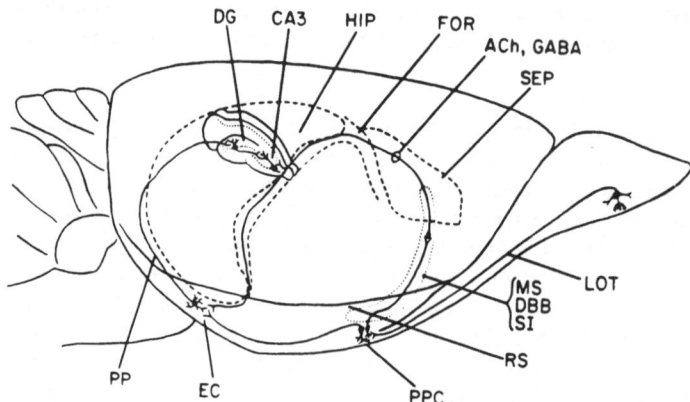

Fig. 13. A schematic side view of the rat brain to show two possible
pathways from the prepyriform cortex (PPC) to the dentate
gyrus (DG). One passes posteriorly through the entorhinal
cortex (EC), as also shown in Fig. 5. The other passes
anteriorly to the medial septum (MS), diagonal band of Broca
(DBB), and substantia innominata (SI). A pathway with both
cholinergic (ACh) and GABAergic fibers originates in the
MS/DBB complex and travels to the DG and EC. See text.

cortex were greatly increased after lindane. After administration of
lindane, population spikes appeared in this response, whereas they
did not during the pretreatment period. This is in agreement with
the appearance of additional population spikes in the response evoked
in CA3 by stimulation of the dentate gyrus after administration of
lindane. In addition, the duration of the increased amplitude of the
response evoked in the dentate gyrus by stimulation of the prepyriform cortex after administration of either lindane or dieldrin is
longer than the reported half-life of these compounds in brain which
has been found to be 24 hr or less (Joy, 1985) or 1.5 days (Vohland
et al., 1981) for lindane and three days (Robinson et al., 1969) for
dieldrin in the rat. The increased amplitude of this response,
produced by a dose of 30 mg/kg of lindane, lasted longer than effects
of the higher dose on overt seizures, hypothermia and anorexia.

The increased amplitude of this response may be attributed
either to a presynaptic effect resulting in increased neurotransmitter release or to a postsynaptic effect resulting in increased
responsiveness of the postsynaptic neuron to the released neurotransmitter. Increased neurotransmitter release would be consistent
with previous studies of the mechanisms of action of lindane and
could be explained by increased calcium entry into presynaptic terminals (see earlier section). Theoretically, reduced presynaptic
inhibition could also increase neurotransmitter release.

The long duration of the increased amplitude of the response evoked in the dentate gyrus by stimulation of the prepyriform cortex has been called an example of long-term potentiation (LTP) by us (Woolley et al., 1984, 1985). LTP is defined as a long-lasting, stable increase in synaptic efficacy. Various limbic system pathways have been shown to be particularly susceptible to demonstrating LTP. LTP in the limbic system, particularly in the hippocampal formation, has been proposed to form the basis of memory (reviewed by Eccles, 1983; Voronin, 1983; Woolley et al., 1984). We have proposed that LTP, at least in the PPC-evoked DG response, is basic to kindling and is associated with enhanced susceptibility to seizures (Chandler et al., 1982; Woolley et al., 1984). Thus, lindane produces both a proconvulsant state (Hulth et al., 1976; Joy et al., 1982) and LTP in the response evoked in the dentate gyrus by stimulation of the pre-pyriform cortex. The mechanisms underlying LTP in the limbic system are currently under intense investigation and dispute (Swanson et al., 1982; Eccles, 1983; Voronin, 1983). Excessive calcium entry, into either presynaptic or postsynaptic neurons or both, may play a key role in the formation of LTP. High levels of extracellular calcium are required for the formation of LTP in the hippocampal slice. It has been hypothesized that excessive calcium entry may produce secondary changes, perhaps involving protein phosphorylation, increased neurotransmitter receptors, or increased neurotransmitter release (Eccles, 1983; Voronin, 1983).

The fact that the amplitude of the response of the dentate gyrus to stimulation of the prepyriform cortex was dramatically increased after administration of lindane, whereas the response of the same structure to stimulation of the perforant path was changed very little by lindane, deserves additional consideration. The pathway between the prepyriform cortex and dentate gyrus involves a synapse in the lateral entorhinal cortex and the perforant path serves as the final input to the outer dendritic layers of the dentate granule cells, as explained previously. Two possible explanations may be presented to account for the fact that the input from the prepyriform cortex results in potentiation of the dentate response whereas the more direct monosynaptic input from the perforant path does not. First, the potentiation may actually occur in the lateral entorhinal cortex, rather than in the dentate gyrus. Second, stimulation of the prepyriform cortex may activate an alternate path through the dia-gonal band/medial septum to the dentate gyrus and entorhinal cortex (Fig. 13), in addition to the path through the lateral entorhinal cortex and perforant path. The first hypothesis is not as easy to test as one might assume, because neurons in the lateral entorhinal cortex are not organized into layers ideally suited for studies of evoked potentials. Evidence for the second hypothesis rests on the fact that stimulation of the prepyriform cortex synaptically acti-vates neurons in the region of the substantia innominata/horizontal limb of the diagonal band of Broca in the rat (Moyano and Molina, 1982) and cat (Yoshihara and DeFrance, 1976). In turn, neurons of the diagonal band/medial septal complex provide both cholinergic and GABAergic projections to the hippocampal formation (Fig. 13). The cholinergic projection has been known for some time (Lewis et al., 1967; reviewed by Swanson et al., 1982), whereas the GABAergic input has been described only recently (Kohler et al., 1984). Stimulation of the medial septal/diagonal band complex may potentiate the input to the dentate gyrus from the perforant path (Fantie and Goddard, 1982). It has been suggested that neurons of the diagonal band provide a route by which the prepyriform cortex can influence the hippocampal formation (Yoshihara and DeFrance, 1976). Thus, stimulation of the prepyriform cortex may simultaneously activate one pathway to the dentate gyrus via the diagonal band and another pathway via the lateral entorhinal cortex and perforant path. Lindane may preferentially affect the pathway through the diagonal band, perhaps the GABAergic component, and in this way potentiate the input from the other path.

Does Lindane Increase Responsiveness of the Postsynaptic Site?

To date, there is little evidence that lindane enhances post-synaptic responsiveness to neurotransmitter, although Joy (1982) has presented electrophysiological evidence that dieldrin has such a postsynaptic effect in the motor cortex upon somatosensory stimulation in the cat. In the present study, lindane increased the firing of postsynaptic neurons in the dentate gyrus and CA3 when the perforant path and dentate gyrus were stimulated, respectively. This was shown either by increased amplitude or occurrence of population spikes in the absence of a change in the amplitude of the population EPSPs. If the amplitude of the population EPSPs is proportional to the neurotransmitter released in the dendrites, then the increased occurrence or amplitude of the population spikes suggests that increased firing of the neurons occurred in the absence of increased neurotransmitter release. In other words, there was enhanced excitability of the postsynaptic neuron. However, this effect was slight, especially in comparison with the increased amplitude of the response evoked in the dentate gyrus by stimulation of the prepyriform cortex. Certainly more work is required before a possible effect of lindane on such postsynaptic sites can be evaluated more carefully.

Insecticides and Body Temperature

Considering the rapidity, magnitude and duration of the hypothermia produced by lindane, it is surprising that the phenomenon has not received attention previously. In other work in this laboratory (Woolley, Griffin and Glass, unpublished observations), it was found that lindane injected intraperitoneally (7.5 mg/kg body weight dissolved in DMSO) significantly reduced colonic temperatures more than 2 degrees C in 20 minutes (the earliest time at which temperatures were measured) and temperatures remained significantly reduced for several hours. A dose of 5 mg/kg via the same route and vehicle also reduced body temperatures, even though no overt seizures occurred. Because hypothermia of such magnitude can affect other endpoints, e.g., evoked potentials, the present findings, indicate that body temperatures should be monitored in studies of the effects of lindane. (In the present studies of the effects of lindane on evoked potentials, colonic temperatures were not affected by 30 mg/kg lindane in oil po.) The fact that the hypothermia could be prevented by diazepam and phenobarbital suggests that enhancement of GABAergic transmission could prevent the effect. Indeed, a GABAergic component in hypothalamic temperature regulation has been described (Bligh, 1981). However, thermoregulation is complex and the mechanisms involved in lindane-induced hypothermia deserve further study.

Because we had previously observed that dieldrin lowered colonic temperatures in the rat (Table 1; Swanson and Woolley, 1982; Woolley et al., 1985) and because dieldrin and lindane have many similar effects and mechanisms of actions, the lindane-induced hypothermia was not unexpected. However, the lindane-induced hypothermia was more severe and longer-lasting than that produced by dieldrin. On the other hand, the duration of seizures was longer in dieldrin-treated rats than in rats receiving a similar dose of lindane (Woolley et al., 1985; Woolley and Dodge, unpublished observations). Wagner and Greene (1978) investigated the time course of the hypothermia produced in the rat by dieldrin and tried to relate hypothermia to brain levels of serotonin, dopamine and norepinephrine. Hypothermia was maximal about 2-3 hr after administration of dieldrin, and levels of both norepinephrine and serotonin, but not dopamine, were reduced in whole brain by acute dieldrin treatment. Hrdina et al. (1974b) may have been the first to point out that a number of cyclodiene insecticides (alpha-chlordane, endrin, heptachlor and heptachlor epoxide) produced hypothermia in the rat. Administration of alpha-methyl-p-tyrosine to block synthesis of the catecholamines partially antagonized the hypothermia produced by alpha-chlordane

Table 1. Effects of insecticides on body temperature in the rat

Insecticide	Single Dose* (mg/kg)	Colonic Temperature	Duration
p,p´- DDT	600	↑	2-4 hr before death
Chlordecone	55 or 75	↓	3-24 hr (transient)
Dieldrin	40	↓	2-4 hr (transient)
Lindane	40 or 50	↓	from 45 min to 3 days (immediate and long-lasting)

*Given in oil po. Studies by Woolley and colleagues. See text.

(Hrdina et al., 1974a,b, 1975). We recently pointed out that another organochlorine insecticide, chlordecone, also produced hypothermia (Table 1; Swanson and Woolley, 1982). Furthermore, it has been known for some time that the anticholinesterase insecticides which cross the blood-brain barrier evoke a striking hypothermia in the rat (Meeter and Wolthuis, 1968), as do other cholinergic agents (Lin et al., 1979).

Thus, a surprisingly large number of insecticides produce hypothermia -- a fact which is probably not generally appreciated in most studies of the toxicological effects of these agents. On the other hand, one insecticide -- DDT -- produces hyperthermia (Table 1), as first pointed out by Henderson and Woolley (1970) and Woolley (1973). DDT-induced hyperthermia was attributed to increased serotonin levels or turnover (Hrdina et al., 1975; reviewed by Woolley, 1982). Considering the great strides made in understanding the role of additional neurotransmitters, including the peptide neuromodulators, in temperature regulation, it seems that it is time to reevaluate the mechanisms by which many of the insecticides alter temperature regulation.

Insecticides and Food Intake

The marked anorexia produced by lindane in the rat in the present study agrees with the finding that topical application of lindane produced severe anorexia in weanling rabbits and some anorexia in young adult rabbits (Hanig et al., 1976). It is also consistent with clinical reports that low doses of an oral lindane emulsion produce nausea (Graeve and Herrnring, 1951). Similarly, nausea was a common complaint of workers exposed to dieldrin while spraying it in fields (Hayes, 1957, 1959). In other studies in this laboratory, we have noted that a 30 mg/kg dose of lindane in oil po significantly reduced food intake for 3 days but did not affect body temperature (Table 2; Zimmer and Woolley, unpublished observations). Thus, the lindane-induced anorexia has a lower threshold than that for the hypothermia. Similarly, whereas a dose of 40 mg/kg of dieldrin in oil po produced only transient effects on body temperature (Table 1; Woolley et al., 1985), the anorexia it produced lasted for days (Table 2; Woolley and Dodge, unpublished observations). However, of the organochlorine insecticides tested in our laboratory, the one which produced the most prolonged reduction in food intake was chlordecone (Table 2; Swanson and Woolley, 1982).

Diazepam and phenobarbital blocked both the lindane-induced anorexia (present study) and the dieldrin-induced anorexia (Woolley and Dodge, unpublished observations). Examples of drugs and putative neurotransmitters which affect food intake (Table 3) show that several of the agents which increase food intake also enhance GABAergic transmission. Opioids also increase food intake. Since chlordecone reduces hypothalamic beta-endorphin levels in the rat

Table 2. Duration of effects of insecticides in decreasing food
intake

Insecticide	Single Dose* (mg/kg)	Duration
Chlordecone	35, 55 or 75	3 weeks
Dieldrin	40	4 days
Lindane	30, 40 or 50	3-5 days

*Administered in oil po. Studies by Woolley and
colleagues. See text.

(Ali et al., 1982; Hong and Ali, 1982), it may be that reduced
endogenous opioids contribute to the chlordecone-induced reduced food
intake.

In view of the recent progress in understanding some of the
complexities of the regulation of food intake and the role that the
traditional and newer putative neurotransmitters play (Morley et al.,
1983; Blundell, 1984), it may be time to make a systematic analysis

Table 3. Agents affecting food intake

A. Examples of Drugs and Putative Neurotransmitters which
Decrease Food Intake.

Naloxone (mu opioid receptor blocker)
Cholecystokinin (CCK; may lead to aversion)
Other peptides (bombesin, enterogastrone, somatostatin)
Serotonin (excites satiety system)
Dopamine (inhibits feeding system)
β-adrenergic stimulation (inhibits feeding system)
Atropine

B. Examples of Drugs and Putative Neurotransmitters which
Increase Food Intake.

Barbiturates
Benzodiazepines (diazepam, chlordiazepoxide)
Muscimol (GABA agonist; ICV)
Opioids (β-endorphin, enkephalin analogs, dynorphin; lCV)
Proglumide (CCK receptor blocker)
Formamidines
Serotonin antagonists and synthesis inhibitors
Norepinephrine (alpha inhibition of satiety system)

Based on Blundell (1984) and Morley et al. (1983). ICV =
intracerebroventricular administration.

of factors contributing to the anorexia produced by a number of
organochlorine and other insecticides.

REFERENCES

Abalis, I.M., Eldefrawi, M.E., and Eldefrawi, A.T., 1985, High
    affinity stereospecific binding of cyclodiene insecticides and
    gamma-BHC to GABA receptors in rat brain, Pestic. Biochem.
    Physiol., in press.
Ali, S.F., Hong, J.-S., Wilson, W.E., Lamb, J.C., Moore, J.A., Mason,
    G.A., and Bondy, S.C., 1982, Subchronic dietary exposure of
    rats to chlordecone (Kepone) modifies levels of hypothalamic
    beta-endorphin, NeuroToxicology, 3(2):119-124.
Andersen, P., Holmqvist, B., and Voorhoeve, P., 1966, Entorhinal
    activation of dentate granule cells, Acta Physiol. Scand.,
    66:448-460.
Bligh, J., 1981, Amino acids as central synaptic transmitters or
    modulators in mammalian thermoregulation, Fed. Proc., 40:2746-
    2749.
Blundell, J.E., 1984, Systems and interactions: an approach to the
    pharmacology of eating and hunger, in: "Eating and Its
    Disorders," Stunkard, A.J. and Stellar, E., eds., Raven Press,
    New York.
Cattabeni, F., Pastorello, M.C., and Eli, M., 1983, Convulsions
    induced by lindane and the involvement of the GABAergic system,
    in: "Toxicology in the Use, Misuse and Abuse of Food, Drugs and
    Chemicals," Arch. Toxicol., Suppl. 6:244-249.
Chandler, D.B., Woolley, D.E., and Overmann, S.R., 1982, Amygdaloid
    kindling progressively increases posttetanic potentiation and
    long-term potentiation of responses evoked in the hippocampus
    and dentate gyrus, Proc. West. Pharmacol. Soc., 25:443-447.
Davies, J.E., Dedhia, H.V., Morgade, C., Barquet, A., and Maibach,
    H., 1983, Lindane poisonings, Arch. Dermatol., 119:142-144.
Eccles, J.C., 1983, Calcium in long-term potentiation as a model for
    memory, Neuroscience, 10:1071-1081.
Eldefrawi, M.E., Sherby, S.M., Abalis, I.M., and Eldefrawi, A.T.,
    1985, Interactions of pyrethroid and cyclodiene insecticides
    with nicotinic acetylcholine and GABA receptors, NeuroToxi-
    cology, 6(2):47-62.
Fantie, B.D., and Goddard, G.V., 1982, Septal modulation of the
    population spike in the fascia dentata produced by perforant
    path stimulation in the rat, Brain Res., 252:227-237.
Ghiasuddin, S.M., and Matsumura, F., 1982, Inhibition of gamma-
    aminobutyric acid (GABA)-induced chloride uptake by gamma-BHC
    and heptachlor epoxide, Comp. Biochem. Physiol., 73C:141-144.
Graeve, K., and Herrnring, G., 1951, Uber die toxizitat des gamma-
    hexachlorocyclohexan, Arch. Internat. Pharmakodyn., 85:64-72.
Halpern, L.K., Wooldridge, W.E., and Weiss, R.S., 1950, Appraisal of
    the toxicity of the gamma isomer of hexachlorocyclohexane in

clinical usage, Arch. Dermatol., 62:648-650.

Hanig, J.P., Yoder, P.D., and Krop, S., 1976, Convulsions in wean-
    ling rabbits after a single topical application of 1% lindane,
    Toxicol. Appl. Pharmacol., 38:463-469.

Hasan, Z., Zimmer, L., and Woolley, D., 1984, Time course of the
    effects of trimethyltin on limbic evoked potentials and distri-
    bution of tin in blood and brain in the rat, NeuroToxicology,
    5(2):217-244.

Hayes, W.J., Jr., 1957, Dieldrin poisoning in man, Pub. Health
    Reports, 72:1087-1091.

Hayes, W.J., Jr., 1959, The toxicity of dieldrin to man. Report of
    a survey, Bull. World Health Org., 20:891-912.

Henderson, G.L., and Woolley, D.E., 1970, Mechanisms of neurotoxic
    action of 1,1,1-trichloro-2,2-bis(p-chlorophenyl)ethane (DDT) in
    immature and adult rats, J. Pharmacol. Exp. Therap., 175:60-68.

Herbst, M., and Bodenstein, G., 1972, Toxicology of lindane, in:
    "Lindane. Monograph of an Insecticide," E. Ulmann, ed., Verlag
    K. Schillinger, Freiburg im Breisgau, W. Germany.

Hill, D.R., and Bowery, N.G., 1981, 3H-Baclofen and 3H-GABA bind to
    bicuculline-insensitive $GABA_b$ sites in rat brain, Nature,
    290:149-152.

Hong, J.-S., and Ali, S.F., 1982, Chlordecone (Kepone) exposure in
    the neonate selectively alters brain and pituitary endorphin
    levels in prepuberal and adult rats, NeuroToxicology., 3(2):111-
    118.

Hrdina, P., Peters, D.A.V., and Singhal, R.L., 1974a, Role of
    noradrenaline, 5-hydroxytryptamine and acetylcholine in the
    hypothermic and convulsive effects of α-chlordane in rats, Eur.
    J. Pharmacol., 26:306-312.

Hrdina, P.D., Singhal, R.L., and Ling, G.M., 1975, DDT and related
    chlorinated hydrocarbon insecticides: Pharmacological basis of
    their toxicity in mammals, Adv. Pharmacol. Chemother., 12:31-88.

Hrdina, P.D., Singhal, R.L. and Peters, D.A.V., 1974b, Changes in
    brain biogenic amines and body temperature after cyclodiene
    insecticides, Toxicol. Appl. Pharmacol., 29:119 (abstract).

Hulth, L., Larsson, M., Carlsson, R., and Kihlstrom, J.E., 1976,
    Convulsive action of small single oral doses of the insecticide
    lindane, Bull. Environ. Contam. Toxicol., 16:133-137.

Joy, R.M., 1982, Mode of action of lindane, dieldrin and related
    insecticides in the central nervous system, Neurobehav. Toxicol.
    Teratol., 4:813-823.

Joy, R.M., 1985, The effects of neurotoxicants on kindling and
    kindled seizures, Fundam. Appl. Toxicol., 5:41-65.

Joy, R.M. and Albertson, T.E., 1984, Lindane and limbic excit-
    ability, NeuroToxicology, 5(4):73 (abstract).

Joy, R.M., Stark, L.G., Peterson, S.L., Bowyer, J.F., and Albertson,
    T.E., 1980, The kindled seizure: production of and modifica-
    tion by dieldrin in rats, Neurobehav. Toxicol., 2:117-124.

Joy, R.M., Stark, L.G., and Albertson, T.E., 1982, Proconvulsant
    effects of lindane: enhancement of amygdaloid kindling in the

rat, <u>Neurobehav. Toxicol. Teratol.</u>, 4:347-354.

Joy, R.M., Stark, L.G., and Albertson, T.E., 1983, Proconvulsant actions of lindane: effects on afterdischarge thresholds and durations during amygdaloid kindling in rats, <u>NeuroToxicology</u>, 4:211-220.

Kadous, A.A., Ghiasuddin, S.M., Matsumura, F., Scott, J.G., and Tanaka, K., 1983, Difference in the picrotoxinin receptor between the cyclodiene-resistant and susceptible strains of the German cockroach, <u>Pestic. Biochem. Physiol.</u>, 19:157-166.

Katz, B., and Miledi, R., 1970, Further study on the role of calcium in synaptic transmission, <u>J. Physiol. (Lond.)</u>, 207:789-801.

Khare, S.B., Rizvi, A.G., Shukla, O.P., Singh, R.P., Perkash, O., Misra, V.D., Gupta, J.P., and Sethi, P.K., 1977, Epidemic outbreak of neuro-ocular manifestations due to chronic BHC poisoning, <u>J. Assoc. Physicians India</u>, 25:215-222.

Kohler, C., Chan-Palay, V., and Wu, J.-Y., 1984, Septal neurons containing glutamic acid decarboxylase immunoreactivity project to the hippocampal region in the rat brain, <u>Anat. Embryol.</u>, 169:41-44.

Lawrence, L.J., and Casida, J.E., 1984, Interactions of lindane, toxaphene and cyclodienes with brain-specific <u>t</u>-butylbicyclophosphorothionate receptor, <u>Life Sci.</u>, 35:171-178.

Lewis, P.R., Shute, C.C.D., and Silver, A., 1967, Confirmation from choline acetylase analyses of a massive cholinergic innervation to the rat hippocampus, <u>J. Physiol. (Lond.)</u>, 191:215-224.

Lievremont, M., Barnier, J.V., and Potus, J., 1984, Gamma-hexachlorocyclohexane inhibition of the calcium fluxes at the desensitized mouse neuromuscular junction, <u>Toxicol. Appl. Pharmacol.</u>, 76:280-287.

Lin, M.T., Chen, F.F., Chern, Y.F., and Fung, T.C., 1979, The role of the cholinergic system in the central control of thermo-regulation in rats, <u>Can. J. Physiol. Pharmacol.</u>, 57:1205-1212.

Lomo, T., 1971, Patterns of activation in a monosynaptic cortical pathway: the perforant path input to the dentate area of the hippocampal formation, <u>Exp. Brain Res.</u>, 12:18-45.

Matsumura, F., and Ghiasuddin, S.M., 1983, Evidence for similarities between cyclodiene type insecticides and picrotoxinin in their action mechanisms, <u>J. Environ. Sci. Health</u>, B18:1-14.

Matsumura, F., and Tanaka, K., 1984, Molecular basis of neuroexcita-tory actions of cyclodiene-type insecticides, <u>in</u>: "Cellular and Molecular Toxicology," T. Narahashi, ed., Raven Press, New York.

McNamara, J.O., 1984, Role of neurotransmitters in seizure mechanisms in the kindling model of epilepsy, <u>Fed. Proc.</u>, 43:2516-2520.

McNaughton, B.L., and Barnes, C.A., 1977, Physiological identifica-tion and analysis of dentate granule cell responses to stimula-tion of the medial and lateral perforant pathways in the rat, <u>J. Comp. Neur.</u>, 175:439-454.

Meeter, E., and Wolthuis, O.L., 1968, The effects of cholinesterase inhibitors on the body temperature of the rat, <u>Eur. J. Pharmacol.</u>, 4:18-24.

Morley, J.E., Levine, A.S., Yim, G.K., and Lowy, M.T., 1983, Opioid modulation of appetite, Neurosci. Biobehav. Rev., 7:281-305.

Moyano, H.F., and Molina, J.C., 1982, Olfactory connections of substantia innominata and nucleus of the horizontal limb of the diagonal band in the rat: an electrophysiological study, Neurosci. Lett., 34:241-246.

Narahashi, T., 1971, Effects of insecticides on excitable tissues, Adv. Insect. Physiol., 8:1-93.

Narbonne, P., and Lievremont, M., 1983, Increase of synaptosomal calcium uptake by lindane in vitro., C.R. Acad. Sci. Paris, 296(series III):811-814.

Normann, T.C., and Samaranayaka-Ramasamy, M., 1977, Secretory hyperactivity and mitochondrial changes in neurosecretory cells of an insect. Cellular effects of the insecticide lindane, Cell Tiss. Res., 183:61-69.

Olsen, R.W., 1982, Drug interactions at the GABA receptor-ionophore complex, Annu. Rev. Pharmacol. Toxicol., 22:245-277.

Olsen, R.W., Wong, E.H.F., Stauber, G.B., and King, R.G., 1984, Biochemical pharmacology of the gamma-aminobutyric acid receptor/ionophore protein, Fed. Proc., 43:2773-2778.

Publicover, S.J., and Duncan, C.J., 1979, The action of lindane in accelerating the spontaneous release of transmitter at the frog neuromuscular junction, Naunyn-Schmiedeberg's Arch. Pharmacol., 381:179-182.

Racine, R., 1978, Kindling: the first decade, Neurosurgery, 3:234-252.

Robinson, J., Roberts, M., Baldwin, M., and Walker, A.I.T., 1969, The pharmacokinetics of HEOD (dieldrin) in the rat, Fd. Cosmet. Toxicol., 7:317-332.

Seifert, J., and Casida, J.E., 1985, Solubilization and detergent effects on interactions of some drugs and insecticides with the t-butylcyclophosphorothionate binding site within the gamma-aminobutyric acid receptor-ionophore complex, J. Neurochem., 44:110-116.

Shankland, D.L., and Schroeder, M.E., 1973, Pharmacological evidence for a discrete neurotoxic action of dieldrin (HEOD) in the American cockroach, Periplaneta americana (L)., Pestic. Biochem. Physiol., 3:77-85.

Solomon, L.W., Fahrner, L., and West, D.P., 1977, Gamma benzene hexachloride toxicity, Arch. Dermatol., 113:353-357.

Stark, L.G., Joy, R.M. and Albertson, T.E., 1983, The persistence of kindled amygdaloid seizures in rats exposed to lindane, Neuro-Toxicology, 4(2):221-226.

Storm-Mathisen, J., 1977, Localization of transmitter candidates in the brain: the hippocampal formation as a model, Prog. Neuro-biol., 8:119-181.

Swanson, L.W., Teyler, T.J., and Thompson, R.F., 1982, Hippocampal long-term potentiation: mechanisms and implications for memory, Neurosci. Res. Prog. Bull., 20:611-769.

Swanson, K.L., and Woolley, D.E., 1980, Dieldrin induced changes in hippocampal evoked potentials in the rat, Proc. West. Pharmacol., 23:81-84.

Swanson, K.L. and Woolley, D.E., 1982, Comparison of the neurotoxic effects of chlordecone and dieldrin in the rat, NeuroToxicology, 3(2):81-102.

Trifiletti, R.R., Snowman, A.M., and Snyder, S.H., 1984, Barbiturate recognition site on the GABA benzodiazepine receptor complex is distinct from the picrotoxinin/TBPS recognition site, Eur. J. Pharmacol., 106:441-448.

Uchida, M., Irie, Y., Fujita, T., and Nakajima, M., 1975, Effects of Nereistoxin on the neuroexcitatory action of insecticides, Pestic. Biochem. Physiol., 5:253-257.

Ulmann, E., 1972, Lindane-Monograph of an Insecticide, Freiburg im Breisgau, West Germany, Verlag K. Schillinger, 383 pages.

Vohland, H.W., Portig, J., and Stein, K., 1981, Neuropharmacological effects of isomers of hexachlorocyclohexane. 1. Protection against pentylenetetrazol-induced convulsions, Toxicol. Appl. Pharmacol., 57:425-438.

Voronin, L.L., 1983, Long-term potentiation in the hippocampus, Neuroscience, 10:1051-1069.

Wagner, S.R. and Greene, F.E., 1978, Dieldrin-induced alterations in biogenic amine content of rat brain, Toxicol. Appl. Pharmacol., 43:45-55.

Wang, C.M., Narahashi, T., and Yamada, M., 1971, The neurotoxic action of dieldrin and its derivatives in the cockroach, Pestic. Biochem. Physiol., 1:84-91.

Wilson, R.C., and Steward, O., 1978, Polysynaptic activation of the dentate gyrus of the hippocampal formation: an olfactory input via the lateral entorhinal cortex, Exp. Brain Res., 33:523-534.

Winteringham, F.P.W., and Barnes, J.M., 1955, Comparative response of insects and mammals to certain halogenated hydrocarbons used as insecticides, Physiol. Rev., 35:701-739.

Wooldridge, W.E., 1948, The gamma isomer of hexachlorcyclohexane in the treatment of scabies, J. Invest. Dermatol., 10:363-366.

Woolley, D.E., 1973, Studies on 1,1,1-trichloro-2,2-bis(p-chlorophenyl)ethane (DDT)-induced hyperthermia: effects of cold exposure and aminopyrine injections, J. Pharmacol. Exp. Therap., 184:261-268.

Woolley, D.E., 1977, Electrophysiological techniques in toxicology, in: "Behavioral Toxicology: An Emerging Discipline," L.W. Reiter, and H. Zenick, eds., EPA Publication No. 600/9-77-042.

Woolley, D.E., 1982, Neurotoxicity of DDT and possible mechanisms of action, in: "Mechanisms of Actions of Neurotoxic Substances," Prasad, K.N., and Vernadakis, A., eds., Raven Press, New York.

Woolley, D., and Zimmer, L., 1984, Effects of organochlorine insecticides: unsolved problems, NeuroToxicology, 5(4):72 (abstract).

Woolley, D., Zimmer, L., Dodge, D., and Swanson, K., 1985, Effects of lindane-type insecticides in mammals: unsolved problems, NeuroToxicology, 6(2):165-192.

Woolley, D., Zimmer, L., Hasan, Z., and Swanson, K., 1984, Do some
    insecticides and heavy metals produce long-term potentiation in
    the limbic system?, in: "Cellular and Molecular Neurotoxi-
    cology," T. Narahashi, ed., Raven Press, New York.
Yamasaki, T., and Ishii, T., (Former name of Narahashi, T.), 1954,
    Studies on the mechanism of action of insecticides. X. Nervous
    activity as a factor in development of gamma-BHC symptoms in the
    cockroach, Botyu-Kagaku (Scientific Insect Control), 19:106-112
    (In Japanese).
Yoshihara, H., and DeFrance, J.F., 1976, Deep temporal lobe projec-
    tions to the nucleus of the diagonal band of Broca.,
    Experientia, 32:55-57.
Zimmer, L., and Woolley, D., 1984, Lindane potentiates hippocampal
    potentials evoked by stimulation of the prepyriform cortex
    (PPC), Soc. Neurosci. Abstr., 10:1202 (abstract).

ALTERED PICROTOXININ RECEPTOR AS A CAUSE FOR CYCLODIENE RESISTANCE IN MUSCA DOMESTICA, AEDES AEGYPTI AND BLATTELLA GERMANICA

Keiji Tanaka and Fumio Matsumura

Pesticide Research Center
Michigan State University
East Lansing, Michigan 48824-1311

INTRODUCTION

Recently it has been reported from this research laboratory that the cyclodienes and γ-BHC resistant German cockroach strains are also resistant to picrotoxinin, a naturally occurring neuroexcitant and one of GABA antagonists (Matsumura and Ghiasuddin, 1983; Kadous et al., 1983). Moreover, the components of the central nervous system (CNS) from the cyclodiene-resistant German cockroaches have been found to have lower specific binding capacities toward picrotoxinin than do the susceptible counterparts (Kadous et al., 1983). It has been also shown that cyclodienes, γ-BHC and some of known GABA antagonists bind competitively with the picrotoxinin receptor in the CNS of the American cockroach (Tanaka et al., 1984). Other agents such as toxaphene and bicyclophosphates also bind competitively with the picrotoxinin receptor in the CNS of the American cockroach.

Dieldrin (Shankland and Schroeder, 1973), γ-BHC (Uchida et al., 1976) and picrotoxinin (Tanaka, et al., 1984) have been shown to interact with the presynaptic area of the American cockroach CNS and thereby stimulate the excitatory neurotransmitter release. [As to the cause for such stimulation it has been proposed by this research group that these agents specifically interact with the putative picrotoxinin receptor which is closely related to the chloride ionophore in the γ-aminobutyric acid-chloride ionophore complex at the presynaptic region, and that such interaction causes inhibition of chloride anion uptake which is regulated by GABA to modulate the presynaptic membrane potential (Ghiasuddin and Matsumura, 1982; Matsumura and Ghiasuddin, 1983; Kadous et al., 1983; Tanaka et al., 1984).]

While mounting evidence shows that the picrotoxinin receptor interaction plays a vital role in the action mechanism of γ-BHC and cyclodiene insecticides in the American and the German cockroaches, several questions as to the scope of the picrotoxinin receptor involvement in other insects and chemical agents still remain unanswered. Accordingly, we have set four major objectives to address some of these questions in this project. First, how frequently the picrotoxinin receptor mediated poisoning processes are observed in other insect species poisoned by cyclodienes. Second, how various cyclodiene-resistant insects have altered the picrotoxinin receptor. Third, how specific its action patterns are, whether one can clearly distinguish the picrotoxinin receptor-medicated action from other effects of insecticides and fourth, which of the neuroactive chemicals really act through this mechanism.

EXPERIMENTAL

The insect materials used for the study were two strains of houseflies, Musca domestica (dieldrin resistant strain, CLD and susceptible strain, SBO), which have been supplied by Dr. F.W. Plapp, Texas A & M University, and three strains of mosquito, Aedes aegypti (the dieldrin and DDT-resistant Isla Verde strain, DDT-resistant but dieldrin-susceptible Trinidad strain, and the susceptible strain, Rockefeller) which have been obtained from Dr. George Craig of Notre Dame University. The strains of the German cockroach, Blatella germanica (susceptible strain, CSMA and dieldrin-resistant strain, LPP), and the American cockroach (Periplaneta americana, L.) are the ones which have been maintained in this laboratory for approximately 20 years.

Toxicity tests

For mosquito larvae, a specific sand coating method was devised as follows: appropriate acetone solutions of picrotoxinin, dieldrin or γ-BHC and piperonyl butoxide were applied to dry sea sand (2g) in a 150 ml beaker. After complete evaporation of the acetone, 100 ml of water were added to the beaker and 20 individuals at 4th instar larvae were released. The mortality was recorded at specific time intervals.

For the housefly bioassay, chemicals were delivered in 0.5 μl of acetone to the thoracic tergum of the housefly (i.e., topical application). The mortality was recorded after 24 hours.

In the case of the German cockroaches, regular insecticides were delivered in 0.5 μl of acetone to the thoracic tergum. α-BHC, 2,10-dichlorbornane, benzodiazepam, SQ-65396, SQ-20009, TETS and t-butylbicyclic phosphate were dissolved in dimethyl sulfoxide and a 0.25 μl aliquot was injected through the abdomen. For some insecticides (pyrethroids, cyclodienes, DDT, chlordimeform and parathion), a film-surface contact method was adopted (Ghiasuddin and Matsumura, 1983).

^{3}H α-Dihydropicrotoxinin (^{3}H-DHPTX) binding assay

The heads of houseflies, American cockroaches or German cockroaches were collected and homogenized in 0.25 M sucrose solution with a glass-glass homogenizer. The subsequent treatments to obtain the crude mitochondrial fraction, and the procedures of the binding assay were identical to the ones reported earlier (Kadous et al., 1983; Tanaka et al., 1984).

Target sensitivity and penetration

The electrophysiological setup used for this study has been described by Scott and Matsumura (1981). The saline used was that of Yamasaki and Narahashi (1959). The spontaneous activity of the isolated abdominal nerve cord of the German cockroach was monitored using a suction electrode placed midway between the fifth and sixth abdominal ganglia. To each preparation, 1 ml of saline with 0.5 µl of ethanol or pesticide in 0.5 µl of ethanol was added twice every five minutes. Symptoms of excitation were defined as four or more "bursts" of typical spontaneous discharge appearing in one minute.

The amount of dihydropicrotoxinin incorporated into the nerve cord of the German cockroach which was set up for electrophysiological measurement was assessed by using ^{3}H-DHPTX. The nerve cord incorporating ^{3}H-DHPTX was solubilized in 0.2 N NaOH, and the radioactivity was measured using a scintillation counter.

Susceptibilities of cyclodiene-resistant houseflies and mosquito larvae to picrotoxinin

Table 1 shows the susceptibilities of a cyclodiene resistant (CLD) and a susceptible (SBO) strain of houseflies to dieldrin, α-BHC and picrotoxinin. It is evident from the data that there is a distinct cross-resistance among these toxicants. This tendency is consistent with the case of mosquito larvae where the dieldrin-resistant strain (Isla Verde) was also found to show cross-resistance to picrotoxinin (Table 2). In the presence of piperonyl butoxide (3 ppm), which was employed to reduce the metabolic detoxification of picrotoxinin, Isla Verde strain was the only strain to show cross-resistance to picrotoxinin. It must be pointed out here that the Trinidad strain is a DDT-resistant but cyclodiene-susceptible strain, indicating that the cross-resistance to picrotoxinin does not extend to other chlorinated hydrocarbon insecticides.

Table 1. Susceptibility of dieldrin-resistant and dieldrin-
          susceptible houseflies to dieldrin, γ-BHC and picrotoxinin

	LD_{50} ($\mu g/\varphi$)[a]		
Strain	Dieldrin[b]	γ-BHC	Picrotoxinin[b,c]
SBO	$12.9 \pm 4.3 \times 10^{-3}$	$3.9 \pm 1.8 \times 10^{-3}$	22.6 ± 6.4
CLD	$254.0 \pm 145.5 \times 10^{-3}$	$22.4 \pm 8.5 \times 10^{-3}$	> 50[d]

[a]LD_{50}after 24 hours (X $\pm$ S.E.).
[b]Topical application on thorax with 0.5 μl of acetone solution.
[c]Piperonyl butoxide (5 μg) was applied, 1 hour prior to the
   application of picrotoxinin.
[d]At 50 μg/fly the actual mortality was 10%.

A similar conclusion was reached by Kadous et al. (1983) who
have shown that a same type of cross-resistance exists among
dieldrin, γ-BHC and picrotoxinin in the German cockroach strains.
Therefore, it appears likely that this phenomenon of the cross-
resistance to picrotoxinin among cyclodiene-resistant insects is
widely distributed among many insect species.

Table 2. Susceptibility of three strains of Aedes aegypti larvae
          against picrotoxinin[a]

			LT_{50}	
Strain	Picrotoxinin	Piperonyl butoxide	Exp. 1	Exp. 2
Trinidad	30(ppm)	3(ppm)	9(hr)	12.5(hr)
Rockefeller	30	3	16	16.5
Isla Verde	30	3	> 48	> 48
Isla Verde	30	9	8	13
Isla Verde	--	9	17.5	18
Isla Verde	30	12	14.3	9
Isla Verde	--	12	9	12

[a]Application method through sand coating. (See Materials and
   Methods).

Penetration of ^{3}H-DHPTX into the nerve cord of the German cockroach

Using an electrophysiological approach, it previously has been shown that the abdominal nerve cord of the cyclodiene-resistant German cockroach is much less sensitive to dieldrin and picrotoxinin than that of the susceptible cockroaches (Kadous et al., 1983). In the current study we have repeated their work using DHPTX instead of picrotoxinin (Table 3) to assess the relationship between the amounts of DHPTX needed to cause the nerve excitation, and the actual amounts of DHPTX which entered into the nerve cord. The results (Table 3) indicate that at 10^{-5} M the nerve cord from the susceptible cockroach shows excitation symptoms earlier than the resistant counterpart. At this dose level, there appears to be a modest, but a significant interstrain difference in the level of ^{3}H-DHPTX pickup between the resistant and the susceptible nerve cords (Fig. 1).

DHPTX is known to be slightly less toxic than picrotoxinin, but still active enough to induce toxicity to mice and houseflies (Miller et al., 1979; Jarboe et al., 1968). At the same time we have also observed that DHPTX induced the same type of spontaneous discharges as picrotoxinin, Y-BHC and dieldrin on the abdominal nerve cord of the German cockroaches as was the case with the American cockroach. As shown in Table 3, the abdominal nerve cord of the resistant cockroach (LPP) is much less sensitive to DHPTX than that of the susceptible one (CSMA). It took 26.5 min. for CSMA nerve cord to develop electrophysiologicallly-typical poisoning symptoms at 10^{-5}M of DHPTX. On the other hand, it took 74.0 min. for LPP nerve cords to reach the same state of excitation under the same experimental conditions. However, there was only a little difference in the amounts of DHPTX taken up by the nerve cords of CSMA and LPP cockroaches. It may be concluded that the sensitivity difference

Table 3. Time to onset of poisoning symptoms in the abdominal nerve cord of susceptible (CSMA) and resistant (LPP) German cockroaches: Dihydro-picrotoxinin (10^{-5}M) with piperonyl butoxide (10^{-4}M)

Strain	Minutes to onset of poisoning symptoms[a]
CSMA	26.5 ± 3.7
LPP	74.0 ± 6.7

[a]Data are expressed as means ± S.E. of five experiments.

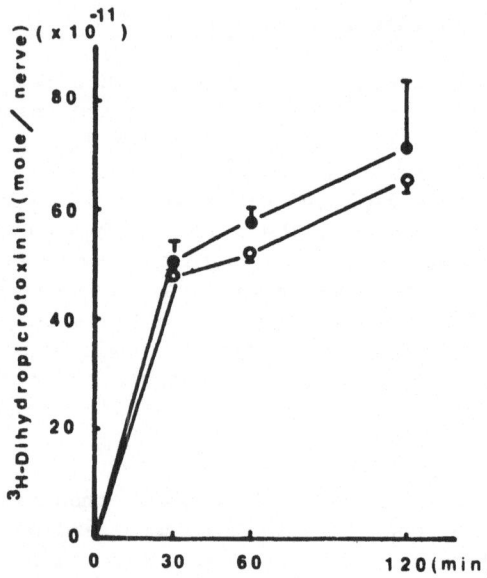

Fig. 1.   Penetration of ^{3}H–DHPTX into the abdominal neve cords of the
two German cockroach strains: dieldrin susceptible CSMA (●)
and resistant LPP (○) strains.   Data are expressed as means
± SE of four nerve cords.

between these two strains comes mainly from the target insensitivity,
not from the penetration difference into nerve cords against
dihydropicrotoxinin.

[3]H–DHPTX binding

i.   Nerve components of houseflies and mosquitoes

The binding of DHPTX to nerve components from houseflies and
mosquitoes was studied as described in the methods section. It has
become evident from the data that the head components of the
resistant housefly strain have less specific binding capacities for
this ligand than the susceptible counterparts (Table 4). The same
tendency was observed in the resistant and the susceptible mosquito
larvae. Among the subcellular fractions from the crude mitochondrial
fraction (Telford and Matsumura, 1970), the fraction containing nerve
ending and some axonic membranes showed the highest specific binding
activity to ^{3}H–DHPTX. An approximate 2–fold difference was found in
the level of ^{3}H–DHPTX specific binding between these two strains
(Table 5).

Table 4. Binding of ^{3}H-DHPTX to a crude mitochondrial fraction from the heads of the resistant and the susceptible strains of the housefly[a]

| | ^{3}H-DHPTX binding | |
Strains	Total (dpm/mg protein)	Specific
Susceptible (SBO)	13399 ± 152	732 ± 203
Resistant (CLD)	12040 ± 78	229 ± 197

[a]The heads from approximately 600 male house flies per experiment were homogenized in 0.25 M sucrose and centrifuged at 800g for 10 min. The supernatant fraction was centrifuged at 20,000g for 45 min and the pellets were resuspended in 0.2 M NaCl in 5 mM sodium phosphate buffer (pH 7.0) containing piperonyl butoxide (10^{-5}) for ^{3}H-DHPTX (specific activity 30Ci/mmol, 11.1 nM final concentration). Data are expressed as means $\pm$ SE of two experiments, each experiment involving three determinations.

ii. Nerve components from the German cockroach heads

The interstrain differences in the picrotoxinin binding receptors from the dieldrin resistant and susceptible German cockroaches were studied in detail. The in vitro binding data on the membrane fraction from their heads were first analyzed through Scatchard plot method (Fig. 2). The interception on the horizontal axis of Scatchard plots has been interpreted to give the information on receptor number (B_{max}). The slope of the line ($-K_d$) indicates the affinity of the receptor to the ligand, where K_d is the dissociation constant of DHPTX to this receptor. As shown in Fig. 2, these two parmeters (B_{max} and K_d) obtained from the heads of resistant and susceptible cockroaches are quite different. The central nervous system of the susceptible strain has about ten times as many receptors as that of the resistant strain. At the same time, the dissociation constant of DHPTX on the receptor from resistant cockroach is one-tenth of that from the susceptible strain.

The difference on the characteristics of picrotoxinin receptor on the resistant and susceptible German cockroaches could also be explained by the different amount of endogenous GABA in their central nervous system (especially presynaptic area on synapse) which could modulate the picrotoxinin receptor to change its characteristics. However, in the presence of 10^{-5}M of GABA, there was little difference in the ^{3}H-DHPTX specific binding to both membrane preparations (Table 6). Therefore, the

Table 5. Binding of ^{3}H–DHPTX to (A) a crude mitochondrial fraction
from the whole bodies of the susceptible and the resistant
strains of the mosquito larvae and (B) a subcellular
fraction from the crude mitochondrial faction[a]

| Strains | ^{3}H-DHPTX[b] binding (dpm/mg protein) | |
	Total	Specific
A) A crude mitochondrial fraction from the whole bodies		
Resistant strain (Isla verde)	50806 ± 89^b	2531 ± 429^b
Susceptible strain (Rockefeller)	47331 ± 131^b	1905 ± 166^b
Susceptible strain (Trinidad)	42492 ± 74^b	2879 ± 85^b
B) A subcellular fraction (fraction of 1.0-1.5 M of sucrose gradient		
Resistant strain (Isla verde)	27123 ± 185^c	1723 ± 338^c
Susceptible strain (Rockefeller)	26261 ± 162^c	2483 ± 129^c
Susceptible strain (Trinidad)	24108 ± 134^c	3034 ± 162^c

[a]Approximately 700 larvae per experiment were used. The
procedure of the preparation of crude mitochondrial fractions
was identical to that shown in Table 4. The subcellular
fraction (the fraction of the sucrose density gradient layer
1.0-1.2M and 1.2-1.5M) was prepared according to the method of
Telford and Matsumura (1970). ^{3}H-DHPTX binding assay was
conducted in the presence of piperonyl butoxide (10^{-5}M).
[b]^{3}H-DHPTX: specific activity 59.6 Ci/mmol, final concentration
16.6 nM.
[c]^{3}H-DHPTX: specific activity 30.0 Ci/mmol, final concentration
11.1 nM.

altered characteristics of the picrotoxinin receptor on the brain
membranes from the resistant cockroach heads are likely due to an
inherent modification of the receptor properties of the PTX
receptor itself.

Characterization of cyclodiene-type actions

It has been debated for a long time what constitutes the class
of cyclodiene-type insecticides (e.g. O'Brien, 1967; Matsumura,
1975). While there is a unanimous agreement that γ-BHC belongs to

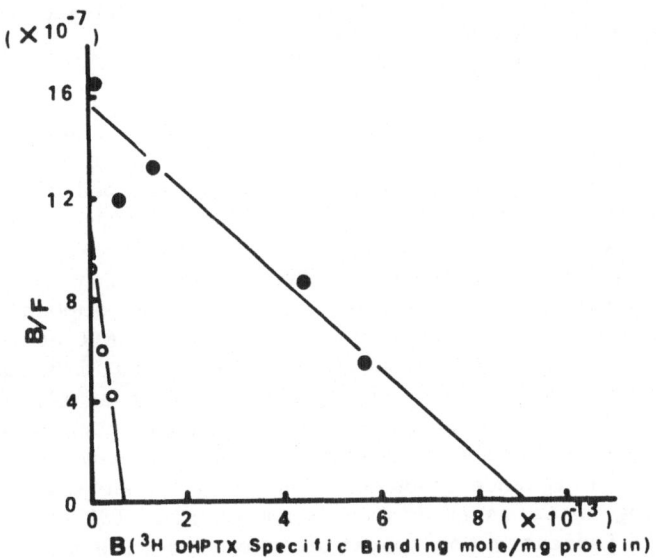

Fig. 2.   Scratchard plot analysis of [^{3}H]-DHPTX binding to the brain
          membrane preparations from two German cockroache strains:
          Dieldrin susceptible CSMA (●) and resistant LPP (○)
          strains. B_{max} receptor number) mole/mg protein was CSMA,
          9.0×10^{-3}; SPP, 7.1×10^{-14}, Kd (dissociation constant
          (M)) was CSMA, 5.8×10^{-7}; LPP, 6.45×10^{-8}. Data are
          expressed as means of two independent experiments, each
          experiment involving three determinations.

Table 6.   Effect of GABA on ^{3}H-DHPTX specific binding[a]

| | ^{3}H-DHPTX[b] specific binding (dpm/mg protein) | |
	CSMA	LPP
None	2489.3 ± 754.4	745.9 ± 291.9
GABA(10^{-5}M)	1998.4 ± 834.3	627.2 ± 313.4

[a]The heads from 110 male cockroaches per experiment were
used. Specific binding was defined as bound
radioactivity to pellet (20,000g, 20 min.) which could
be displaced by 0.1 mM of nonlabeled DHPTX in the
absence of or in the presence of GABA (10^{-5}M). Data
are expressed as means $\pm$ SE of two experiments, each
experiment involving three determinations.
[b]^{3}H-DHPTX (specific activity 30 Ci/mmol, final
concentration 11.1 nM).

this group, the affiliation of toxaphene, mirex and kepone, etc., has not been clearly determined.

To study this question, we have adopted the criteria for any compounds to be qualified for the cyclodiene type insecticides. First, the compound must induce hyperexcitation in the central nervous system of cockroaches in the manner described by several workers (e.g. Shankland and Schroeder, 1973; Uchida et al., 1978). Second, the compound must bind to the picrotoxinin receptor in the central nervous system of insects. Third, the cyclodiene-resistant insects must extend their cross-resistance to the compound in question. The last criterion is important, since all of the studies from this research group (Ghiasuddin and Matsumura, 1982; Kadous et al., 1983; Matsumura and Ghiasuddin, 1983; Tanaka et al., 1984) have shown that the alteration of the picrotoxinin receptor in the resistant German cockroach is specific. The fact that they do not show any cross-resistance even to bicuculline (Kadous et al., 1983), a close homolog in terms of its action mechanism (Olsen et al., 1976), clearly indicates the specificity of its alteration characteristics.

Previously we have shown that cyclodiene insecticides, γ-BHC toxaphene and bicyclic phosphate bind to the picrotoxinin receptor of the central nervous system of the American cockroach (Tanaka et al., 1984). To examine whether the dieldrin-resistant insects show resistance to chemicals shown to bind with the picrotoxinin receptor of the American cockroach, the resistance levels of LPP strain against these chemicals were determined.

As shown in Tables 7, 8 and 9, the LPP strain shows cross-resistance to all cyclodiene insecticides known to interact with the picrotoxinin receptor. The resistance levels are in the order of 2 to 20 fold. In addition, LPP shows resistance to 2, 10-dichlorobronane, γ-BHC and toxaphene. Mirex, though it had been chemically classified as a cyclodiene, was found to be refractory to the picrotoxinin receptor-mediated reactions as judged by the lack of the cockroach responses with regard to cross-resistance and the nerve excitation. Another polychlorinated hydrocarbon insecticide, kepone (= chlordecone), is unusual in several ways. First, it has been shown to have a very high potency of ^{3}H-DHPTX specific binding inhibitor (Tanaka et al., 1984). Second, although at an early stage of poisoning, there was a modest degree of cross-resistance (Table 8); at later stages no significant cross-resistance was observed. Third, its ability to induce dieldrin-type nerve excitation was transient, though it was definitely positive.

Electrophysiologically, dieldrin, γ-BHC and picrotoxinin induce similar and characteristic spontaneous discharge in the cockroach (Narahashi and Yamasaki, 1959; Shankland and Schroeder, 1973, Uchida et al., 1978; Kadous et al., 1983; Tanaka et al., 1984). Among the

Table 7. Comparative toxicities of cyclodiene insecticides to
dieldrin susceptible (CSMA) and resistant (LPP) German
cockroaches[a]

| Cyclodiences[a] | LT_{50}(hours)[b] | | Resistance ratio | Dieldrin[c] type nerve excitation |
	CSMA strain	LPP strain	(LPP/CSMS)	
Aldrin	5.8 ± 0.3	51.1 ± 5.4	8.8 ± 1.2	+[d]
Dieldrin	11.6 ± 0.9	181.6 ± 21.9	15.7 ± 2.2	+
Photodieldrin	11.1 ± 0.5	101.0 ± 8.8	9.1 ± 0.9	+
Heptachlor	4.3 ± 0.4	57.5 ± 7.8	13.4 ± 2.2	+
Heptachlor- epoxide	7.0 ± 0.2	111.2 ± 5.9	15.9 ± 1.0	+
Isodrin	9.1 ± 0.3	21.3 ± 1.7	2.3 ± 0.2	+
Endrin	8.7 ± 0.5	16.3 ± 1.3	1.9 ± 0.2	+
α -Chlordane	14.3 ± 2.1	> 192	> 13.4	+
Oxychlordane	5.3 ± 2.1	60.5 ± 5.4	11.4 ± 1.1	+

[a]Surface contact method (3mg/jar/10 roaches).
[b]Median lethal time, values are given as X ± SE in hours.
[c]At 10^{-5}M concentration of chemicals in 2 hours.
[d]Symbol + means that dieldrin type nerve excitation was observed.

chemicals shown in Tables 7, 8 and 9 DHPTX, toxaphene, α–BHC and t-butylbicyclic phosphate, besides typical cyclodienes, were found to induce the typical and similar spontaneous discharge to dieldrin, γ–BHC and picrotoxinin. TETS failed to cause similar nerve excitation to dieldrin at 10^{-4}M in two hours.

DISCUSSION

From this study, it has become evident that cyclodiene resistant housefly (CLD) and mosquito (Isla Verde) strains have developed resistance to cyclodienes by altering the ligand binding capacity of the target site, i.e., picrotoxinin receptor, in their central nervous system. This conclusion is supported by the findings that these housefly and mosquito strains show cross-resistance to picro-toxinin, and that the membranes of their central nervous systems have lower picrotoxinin binding capacity than the susceptible counter-parts. Kadous et al. (1983) have observed the same phenomenon on the cyclodiene-resistant German cockroaches. Therefore, it appears that this phenomenon of the cyclodiene resistance involving alteration of picrotoxinin receptor, i.e., the target site, may be widely distri-buted among many insect species.

Table 8. Comparative toxicities of various chemicals to dieldrin susceptible (CSMA) and resistant (LPP) German cockroaches[a]

Chemicals	LD_{50}[b] (μg/roach)			Resistance ratio (LPP/CSMA)	Dieldrin[g] type nerve excitation
	Hours	CSMA strain	LPP strain		
Mirex	84	3.12 ± 0.94	2.85 ± 0.95	0.91 ± 0.41	-(10^{-4}M)[h]
Kepone	60	6.89 ± 1.84	14.78 ± 3.86	2.15 ± 0.80	+(10^{-4}M)
Toxaphene	84	5.08 ± 0.83	5.92 ± 1.10	1.17 ± 0.29	+(10^{-5}M)
	24	2.46 ± 0.40	>40[d]	>16.36	
	48	1.28 ± 0.19	>40[e]	>31.25	
α -BHC	24	7.67 ± 4.51	>50[f]	> 6.51	+(10^{-4}M)
γ -BHC	24	0.23 ± 0.07	11.75 ± 2.79	50.87 ± 19.70	+(10^{-5}M)
2,10-dichlorobornane[c]	24	16.90 ± 7.20	55.00 ± 7.80	3.25 ± 1.46	+(10^{-4}M)
Benzodiazepam[c]	24	15.70 ± 7.20	20.90 ± 6.50	1.33 ± 0.74	-(10^{-4}M)
TETS[c]	24	0.32 ± 0.05	1.92 ± 0.31	6.00 ± 1.35	-(10^{-4}M)
t-Butylbicyclic phosphate[c]	24	0.10 ± 0.02	1.46 ± 0.38	14.60 ± 4.79	+(10^{-4}M)
SQ-65396[c]	72	not insecticidal at 30 g			-(10^{-4}M)
SQ-20009[c]	72	not insecticidal at 30 g			-(10^{-4}M)

[a] Topical application method (mirex, toxaphene and γ-BHC) and injection method (α -BHC 2,10-dichlorobornane, benzodiazepam, TETS, t-butylbicyclicphosphate, SQ-65396 and SQ-20009 were used.
[b] Median lethal dose, values are given as X ± SE in μg.
[c] Piperonyl butoxide (10 μg/roach) was applied 1 hour before the treatment of the test chemicals.
[d] No mortality at 40 μg.
[e] 30% mortality at 40 μg.
[f] No mortality at 50 μg.
[g] Symbols + and - mean that dieldrin type nerve excitation was observed and not observed, respectively at the concentration shown in parentheses.
[h] The effect was weak and transient.

Table 9. Comparative toxicities of pyrethroids, parthion, chlordimeform and DDT to dieldrin susceptible (CSMA) and resistant (LPP) German cockroaches[a]

Chemicals	Dose (μg/jar)	KT_{50} (min)[b]		Resistance ratio	Dieldrin[d] type nerve excitation
		CSMA strain	LPP strain	LPP/CSMA	
Bioallethrin	200	8.0 ± 2.3 (26.5 ± 1.6)[c]	5.8 ± 1.4 (26.8 ± 2.0)[c]	0.73 ± 0.27 (1.01 ± 0.08)	$-(10^{-7} M)$
Permethrin	150	18.0 ± 3.4 (26.2 ± 0.7)[c]	17.4 ± 3.4 (22.8 ± 1.0)[c]	0.97 ± 0.26 (0.87 ± 0.04)	$-(10^{-7} M)$
Fenvalerate	150	17.3 ± 3.3 (40.4 ± 8.1)[c]	14.9 ± 2.9 (37.1 ± 3.6)[c]	0.86 ± 0.23 (0.92 ± 0.20)	$-(10^{-7} M)$
Cypermethrin	100	17.3 ± 6.2 (37.5 ± 3.4)[c]	19.6 ± 6.5 (36.2 ± 3.3)[c]	1.13 ± 0.55 (0.97 ± 0.12)	$-(10^{-7} M)$
Deltamethrin	20	13.4 ± 2.7 (35.7 ± 3.9)[c]	9.7 ± 4.5 (33.5 ± 8.2)[c]	0.72 ± 0.37 (0.94 ± 0.25)[c]	$-(10^{-7} M)$
Parathion	1000	65.9 ± 4.3 (7.5 ± 2.0)[c]	58.9 ± 2.1 (8.7 ± 2.9)[c]	0.89 ± 0.07 (1.17 ± 0.50)	$-(10^{-5} M)$
Chlordimeform	1000	126.4 ± 19.2	101.8 ± 16.4	0.81 ± 0.18	$-(10^{-3} M)$
DDT	6000	638.8 ± 107.0	342.6 ± 48.8	0.54 ± 0.12	$-(10^{-5} M)$

[a] Surface contact method (μg/jar/10 roaches).
[b] Median knockdown time, values are given as X ± SE in minutes.
[c] LT_{50} values of the above chemicals (values are given as X ± SE in hours) are shown in prentheses.
[d] Dieldrin type nerve excitation was not observed at the concentration shown in parentheses.

As for mechanisms of resistance to insecticides in general, not only the target site sensitivity, but many other factors could be involved: e.g., decreased penetration of insecticides through the cuticle and penetration into the target site, and increased metabolic detoxication (O'Brien, 1967; Matsumura, 1975; Narahashi, 1982; Georghiou and Saito, 1983). However, in the case of the cyclodience-resistant German cockroach strains, the sensitivity of the target site appears to be the most important factor. Other factors are unlikely to contribute to the occurrence of this type of resistance (see the above reviews).

As to nerve penetration, little interstrain difference was found in the amounts of DHPTX penetrated into the nerve cords, and yet the nerve cord of the susceptible strain showed more sensitivity to DHPTX than the one of the resistant strain. As for metabolic difference, it has been concluded that it does not play an important role on resistance as judged by the isotope effect of lindane-d_6 on its insecticidal activity (i.e. CSMA 3.42 $\pm$ 0.81 and LPP 1.91 $\pm$ 0.78). Lindane-d_6 has been reported to be a good tool for this type of study (Tanaka et al., 1981). The difference of cuticle penetration is also excluded, because the resistance ratio based on LD_{50} values determined by topical application method was found to be the same as that by injection. As a result, the difference in the target site sensitivity may be concluded to be the major cause of cyclodiene resistance.

In mammalian species, the phenomena of differential drug tolerance due to differences in the nature of receptors have been documented, for example, for benzodiazepine, insulin and TCDD, etc. (Rosenberg and Chiu, 1981; Kahn, 1970; Poland and Kende, 1977).

According to the criteria adopted here all cyclodiene insecticides except mirex seem to owe their major insecticidal activities to their ability to interact with the PTX receptor. Two other chlorinated compounds, α-BHC and 2,10-dichlorolornane, probably act in a similar manner, though they are not highly insecticidal. t-Butylbicyclic phosphate definitely acts through the same route.

On the other hand, some of the chemicals other reseachers have found to interact with PTX receptors in mammalian brains were negative in the cross-resistance and the electrophysiological test, indicating that these chemicals are not likely to interact with the PTX receptor from the CNS of the German cockroach. They are various pyrethroids, SQ-65396, SQ-20009, and benzodiazepam (Ticku et al., 1978; Ticku and Olsen, 1979, Leeb-Lundberg and Olsen, 1980; Squires et al., 1983; Lawrence and Casida, 1983). Earlier we have shown that these chemicals do not compete against [3H]-DHPTX at the PTX receptor from the CNS of the American cockroach (Tanaka et al., 1984).

Thus the data obtained from the mammalian CNS definitely differ from those collected from the cockroach CNS. An additional complication is that in mammalian assays some of the workers used [^{35}S]–TBPS as an alternate ligand to [^{3}H]–DHPTX. The former ligand binding is known to be influenced by GABA, (Squires et al., 1983) while the latter is not (Ticku and Olsen, 1978, both tested in the mammalian CNS). Such differences in binding behavior suggest that the binding sites for these two artificial ligands are not exactly identical though there is apparently an overlapping receptor site or sub-populations of receptors reacting to both ligands.

In either case as far as cockroach species are considered it is clear from the results of the current investigation that in terms of insecticidal actions, the involvement of PTX receptor is limited to cyclodiences and few limited structural analogs, and not to others such as pyrethroids, DDT, parathion and chlordimeform.

ACKNOWLEDGEMENTS

Supported by Michigan Agricultural Experiment Station (Journal Article No. 11492), Michigan State University, and by a research grant ES01963 from the National Institute of Environmental Health Sciences, Research Triangle Park, North Carolina. We thank Dr. F.W. Plapp and Dr. George Craig for kindly supplying the resistant and the susceptible houseflies and mosquito strains.

REFERENCES

Georghiou, G.P., and Saito, T., 1983, "Pest Resistance to Pesti-cides," Plenum Press, New York.

Ghiasuddin, S.M., and Matsumura, F., 1982, Inhibition of gamma-aminobutyric acid (GABA)–induced chloride uptake by gamma–BHC and heptachlorepoxide, Comp. Biochem. Physiol., 73:141–144.

Jarboe, E.H., Porter, L.A., and Buckler, R.T., 1968, Structural aspects of picrotoxinin action, J. Med. Chem., 11:729–731.

Kadous, A.A., Ghiasuddin, S.M., Matsumura, F., Scott, J.G. and Tanaka, K., 1983, Difference in the picrotixinin receptor between the cyclodiene-resistant and susceptible strains of the German cockroach, Pestic. Biochem. Physiol., 19:157–166.

Kahn, C.R., 1976, Insulin sensitivity and insulin resistance: regulation of insulin receptor in vivo, in: "Receptors for Viruses, Antigens and Antibodies, Polypeptide Hormones and Small Molecules,", Beers, R.F. and Bassett, E.G., eds., Raven Press, New York.

Lawrence, L.J., and Casida, J.E., 1983, Stereospecific action of pyrethroid insecticides on the γ-aminobutyric acid receptor-ionophore complex, Science, 221:1399–1400.

Leeb-Lundberg, and Olsen, R.W., 1980, Picrotoxinin binding as a probe of the GABA postsynaptic membrane receptor-ionophore complex, in: "Psychopharmacology and Biochemistry of Neurotransmitter Receptors," Yamamura, H.I., Olsen, R.W. and Usdin, E., eds., Elsevier, North Holland, Inc.

Matsumura, F., and Ghiasuddin, S.M., 1983, Evidence for similarities between cyclodiene type insecticides and picrotoxinin in their action mechanism, J. Environ. Sci. Health., B18, 1-14.

Matsumura, F., 1975, "Toxicology of Insecticides," Plenum Press, New York.

Miller, T.A., Maynard, M., and Kennedy, J.M., 1979, Structure and insecticidal activity of picrotoxinin analogs, Pestic. Biochem. Physiol., 10:128-136.

Narahashi, T., 1982, Resistance to insecticide due to reduced sensitivity of the nervous system, in: "Pest Resistance to Pesticides," Georghiou, G., and Saito, T., eds., Plenum Press, New York.

O'Brien, R.D., 1967, "Insecticide: Action and Metabolism," Academic Press, New York.

Olsen, R.W., Ban, M., and Miller, T., 1976, Studies on the neuropharmacological activity of bicuculline and related compounds, Brain Res., 102:283-299.

Poland, A., and Kende, A., 1977, The genetic expression of aryl hydrocarbon hydroxylase activity: Evidence for a receptor mutation in nonresponsive mice, in: "Origins of Human Cancer," Hiatt, H.H., Watson, J.D., Winston, J.A., eds., Cold Spring Harbor Laboratory, Cold Spring Harbor, New York.

Rosenberg, H.C. and Chiu, T.H., 1981, Tolerance during chronic benzodiazepine treatment associated with decreased receptor binding, Eur. J. Pharmacol., 74:453-460.

Scott, J.G. and Matsumura, F., 1981, Characteristics of a DDT-induced case of cross-resistance to permethrin in Blatella germanica. Pestic. Biochem. Physiol., 16:21-27.

Shankland, D.L. and Schroeder, M.E., 1973, Pharmacological evidence for a discrete neurotoxic action of dieldrin (HEOD) in the American cockroach, Periplaneta americana (L.), Pestic. Biochem. Physiol., 3:77-86.

Squires, R.F., Casida, J.E., Richardson, M., and Saederup, E., 1983, (^{35}S) t-butylbicyclophosphorothionate binds with high affinity to brain specific sites coupled to γ-Aminobutyric acid-A and ion recognition sites, Mol. Pharmacol., 23:326-336.

Tanaka, K., Scott, J.G., and Matsumura, F., 1984, Picrotoxinin receptor in the central nervous system of the American cockroach: Its role in the action of cyclodiene-type insecticides, Pestic. Biochem. Physiol., 22:117-127.

Tanaka, K., Nakajima, M., and Kurihara, N., 1981, The mechanism of resistance to lindane and hexadeuterted lindane in the third Yumenoshima strain of housefly, Pestic. Biochem. Physiol., 16:149-157.

Telford, J.N. and Matsumura, F., 1970, Dieldrin binding in subcellu-
    lar nerve components of cockroaches. An electron microscopic
    and autoradiographic study, J. Econ. Entomol., 63:795-800.
Ticku, M.K., Ban, M., and Olsen, R.W., 1978, Binding of (^{3}H) α-
    Dihydropicrotoxinin, a γ-aminobutyric acid synaptic antagonist
    to rat brain membranes, Mol. Pharmacol., 14:391-402.
Ticku, M.K., and Olsen, R.W., 1979, Cage convulsants inhibit picro-
    toxinin binding, Neuropharmacol., 18:315-318.
Uchida, M., Fujita, T., Kurihara, N., and Nakajima, M., 1976,
    Toxicities of γ-BHC and related compounds. in: "Pesticide and
    Venom Neorotoxicity," Shankland, D.L., Hollingworth, R.M. and
    Smyth, T., Jr. eds., Plenum Press, New York.
Yamasaki, T., and Narahashi, T., 1959, The effects of potassium and
    sodium ions on the resting and action potentials of the
    cockroach giant axon, J. Insect Physiol., 3:146-158.

ACETYLCHOLINE AND GABA RECEPTORS IN INSECT CNS AS SITES OF INSECTICIDE ACTION

David B. Sattelle and Sarah C.R. Lummis

AFRC Unit, Department of Zoology,
University of Cambridge,
Downing Street, Cambridge CB2 3EJ, UK

INTRODUCTION

Recent advances in our knowledge of the properties and functions of central nervous system (CNS) neurotransmitters offer new approaches to understanding the molecular mechanisms of insecticide actions. The CNS of insects remains the primary target site for most of the current generation of insecticides and those at present in development (cf. Corbett et al., 1984). Despite rapid progress in the characterization of vertebrate CNS neurotransmitter receptors (Yamamura et al., 1978), work on the properties and functions of insect neurotransmitter receptors is in its infancy.

Neurotransmitter receptors are those membrane proteins which recognize and initiate the neurone's response to the presence of neurotransmitter molecules released from an adjacent cell into the synaptic cleft, a narrow (20nm) space separating pre- and postsynaptic elements of a neuronal pathway. Recent improvements in our understanding of insect CNS neurotransmitter receptors can be attributed to several technical advances in insect neurobiology. These include the use of specific, high-affinity receptor probes which enable radiolabelled ligand binding studies of receptors in membrane fractions derived from insect CNS by subcellular fractionation. A variety of such probes have been successfully employed to characterize vertebrate CNS neurotransmitter receptors (cf. Yamamura et al., 1978). To set in context the recent work on insect CNS acetylcholine and GABA receptors it will be instructive to briefly review advances in receptor studies on vertebrate tissues.

Another significant technical advance that has facilitated progress in insect receptor studies has been the increasing use of

identifiable neurones for electrophysiological experiments on CNS
receptors. The use of single identified cells is a significant
improvement over multifibre preparations and enables detailed interpretation of the pharmacological actions of receptor ligands. The
cells chosen are selected for their size, ease of recognition, their
accessibility for microelectrode experiments, and for their different
functional types – sensory neurones, interneurones, motorneurones and
modulator cells.

A wealth of data is now available on vertebrate neurotransmitter
receptors, many of which are well understood at a molecular level.
There is some evidence that certain insecticides interact with neurotransmitter receptor sites (cf. Corbett et al., 1984). Therefore
detailed characterization of the properties and pharmacological profiles of insect neurotransmitter receptors is a necessary prerequisite to any attempt to understand the actions of insecticidally-
active molecules. In the present paper we first survey some of the
recent advances in the study of acetylcholine and GABA receptors of
vertebrates and then describe recent progress on receptors for these
same neurotransmitters in insect CNS. Finally the interactions of
insecticides with these membrane receptors will be discussed.

REVIEW OF RELATED WORK

Neurotransmitters may act at more than one specific recognition
site. In the case of acetylcholine and GABA distinct subclasses of
receptors can be distinguished in vertebrates using pharmacological
agents that activate the receptor (agonists) and ligands that block
responses to the neurotransmitter (antagonists).

Vertebrate Nicotinic Acetylcholine Receptors

Vertebrate nerve-muscle junctions are the cholinergic synapses
investigated in most detail and a number of the molecular components
of synapses have been purified from this tissue, and the modified
nerve-muscle junctions of electric fish (Electrophorus electricus,
Narke japonica, and species of the Torpedo and Narcine genera). A
total of 5 polypeptide subunits in the probable configuration $\alpha\gamma\alpha\beta\delta$
make up the peripheral nicotinic acetylcholine receptor molecule of
vertebrates though the relative positions of the β and δ subunits
remain uncertain (Popot and Changeux, 1984). All 5 polypeptide
subunits have now been sequenced and each traverses the membrane
several times. The peripheral nicotinic receptor regulates a large
cationic channel 6-7 Å in diameter, though the precise way in which
these subunits line the channel is not known. Channel opening in
response to acetylcholine (ACh) is cooperative, at least two ACh
molecules being required. The snake neurotoxin α-bungarotoxin binds
to sites that recognize ACh with a subnanomolar dissociation constant. The toxin blocks postsynaptically-located receptors in

electrophysiological studies on muscle and electroplax tissue at
nanomolar concentrations. This specific receptor probe has facilitated characterization and purification of the receptor using in
particular Torpedo electric tissue, where the density of nicotinic
receptors reaches 100mg receptor protein per kg wet weight tissue.
The receptor protein has been purified to homogeneity following
solubilization with detergents. Reconstitution experiments show that
this molecule is all that is needed to generate ACh-induced conductance. The best estimates of molecular weight for the functional
peripheral nicotinic receptor of vertebrates are 285,000-290,000.
This receptor is one of the best characterized of all membrane proteins (Popot and Changeux, 1984).

     The success of α-bungarotoxin as a probe for peripheral nico-
tinic ACh receptors has led to its adoption as a receptor probe in
CNS tissues (Morley et al., 1979). However, although specific
binding of α-bungarotoxin has been detected in various CNS tissues of
vertebrates, in several cases nicotinic cholinergic responses are not
blocked by the toxin in physiological experiments on the same tissue.
Studies to date provide evidence for both α-bungarotoxin binding and
physiological block of ACh responses in avian and amphibian autonomic
ganglia, and in the tectum of amphibia and fish (Chiappinelli and
Zigmond, 1978; Oswald and Freeman, 1981; Barnard et al., 1979).
However, in mammalian ganglia specific binding sites for   α-
bungarotoxin are present but they may not be directly related to
nicotinic cholinergic transmission since no block of the nicotinic
ACh response has been reported in this preparation (see Morley et
al., 1979 for review). Differences in dissociation kinetics between
neuronal and peripheral vertebrate nicotinic receptors have been
observed and differences in the pharmacological profiles of CNS and
muscle binding sites are also evident. It is possible that some of
the discrepancies between the actions of α-bungarotoxin at different
neuronal and neuromuscular ACh sites are due to the presence of a
physiologically active contaminant which may mask α-bungarotoxin
inactivity. Recent purification attempts show that although the CNS
nicotinic receptor is a multimeric protein, its subunit composition
may well differ from that of the peripheral receptor (Barnard et al.,
1983).

Vertebrate Muscarinic Acetylcholine Receptors

     Muscarinic ACh receptors form the largest group of cholinergic
receptor types in the vertebrate central nervous system (e.g.
Salvaterra and Foders, 1979). They are linked to a wide variety of
physiological and biochemical responses, both excitatory and inhibitory, in different tissues (Birdsall and Hulme, 1983; Sokolowsky et
al., 1983; Nathanson, 1982). Biochemical signals that have been
demonstrated to be linked to muscarinic receptors include fluxes of
calcium, sodium and potassium, an increase in cyclic guanine nucleotide levels, an increase in phosphatidylinositol turnover and inhibi-

tion of hormone-activated adenylate cyclase (Birdsall and Hulme, 1983; Sokolowsky et al., 1983). Possibly all or some of these responses may be linked, or they may be triggered independently with different systems predominating in different cells or tissues.

Binding studies with radiolabelled antagonists to muscarinic ACh receptors, notably quinuclidinyl benzilate, have been described in a variety of tissues as the interaction of a ligand with a single class of binding sites which have a Hill coefficient close to unity (Birdsall and Hulme, 1976; Kloog and Sokolowsky, 1977, 1978). Muscarinic agonists and antagonists inhibited the binding of these tritiated antagonists whereas drugs with no muscarinic activity did not (Yamamura and Snyder, 1974; Kloog and Sokolowsky, 1978; Bartfai et al., 1976).

More recently it has been observed that the binding characteristics of tritiated antagonists varied between different vertebrate tissues (Birdsall and Hulme, 1983). Such heterogeneity may reflect subclasses of muscarinic receptors, or alternatively may be simply a manifestation of pre-existing interconvertible sites.

Early studies on muscarinic ACh receptors indicated that they were much smaller than nicotinic ACh receptors: Fewtrell and Rang (1973) reported two major protein constituents of smooth muscle muscarinic receptors, with molecular weights of 23,000 and 50,000, while Alberts and Bartfai (1976) reported a molecular weight of 30,000 for the muscarinic ACh receptor from rat brain. Since that time, Birdsall et al. (1979) have solubilized muscarinic ACh receptors from rat, frog and guinea-pig brain and guinea-pig ileum, and the results suggest that the receptor consists of a single polypeptide with a molecular weight of 77,600-83,200. Similarly Haga (1980, 1983), solubilized the muscarinic ACh receptor from rat brain and identified a single polypeptide with a molecular weight of 83,000, as determined from SDS-PAGE analysis, or 86,000 based on gel- filtration experiments.

Vertebrate GABA Receptors

The role of GABA as an inhibitory neurotransmitter in mammalian nervous systems is well documented (Curtis and Johnston, 1974; Krnjevic, 1974; Roberts et al., 1976). Two separate classes of GABA receptor have been described. The $GABA_A$ receptor is associated with a chloride ion channel, while the $GABA_B$ receptor appears to be associated with a change in conductance to calcium and/or potassium ions. It is the $GABA_A$ receptors, or a subclass of these receptors, which are associated with benzodiazepine, barbiturate and avermectin binding sites.

Electrophysiological studies have demonstrated that in vertebrates, $GABA_A$ receptors are linked to a chloride ion channel and activation of the receptor by GABA and $GABA_A$ agonists will produce an influx (hyperpolarization) or efflux (depolarization) of chloride ions (see Gallagher and Shinnick-Gallagher, 1983 for review). The GABA antagonists picrotoxin and bicuculline will selectively block this response (Curtis et al., 1971).

The mechanism of action of $GABA_B$ receptors is not yet fully understood, but there is electrophysiological evidence that $GABA_B$ receptor activation may involve a direct modification of calcium channel conductance (Dunlap and Fischbach, 1981) and/or potassium conductance (Newberry and Nicoll, 1984a,b), possibly the calcium-dependent potassium conductance (Alger and Nicoll, 1980). There is also some evidence that $GABA_B$ receptors may act indirectly via adenylate cyclase (Dolphin, 1984).

Radiolabelled GABA binding studies have been used to examine the characteristics of both the $GABA_A$ and $GABA_B$ binding sites using appropriate conditions to minimize binding to the other site. In addition radiolabelled agonists such as muscimol (predominantly a $GABA_A$ agonist) and baclofen (a $GABA_B$ agonist) have also been used to characterize the receptors (see Olsen, 1982; Bowery et al., 1983 for recent reviews).

EXPERIMENTAL

A great deal of our knowledge of the physiological properties and functions of neurotransmitter receptors in insects stems from findings using identified neurones of the CNS of the cockroach Periplaneta americana. In the following two sections we describe experiments on the ACh and GABA receptors of identified cells.

Acetylcholine Receptors of Identified Insect Neurones

The terminal abdominal ganglion of the cockroach Periplaneta americana contains seven pairs of identified giant interneurones. Cobalt staining of each of these cells reveals their characteristic morphology (See Fig. 1). Cell body position, neurite shape, dentritic branching pattern and the presence or absence of an axon collateral provide anatomical criteria for the identification of these neurones (Harrow et al., 1980). They receive excitatory, monosynaptic input from mechanoreceptor afferents via cercal nerve XI and are highly sensitive to ionophoretically-applied ACh (Callec, 1974; Sattelle, 1980).

α-Bungarotoxin, a vertebrate nicotinic receptor probe, blocks excitatory postsynaptic potentials recorded from cercal afferent, giant interneurone synapses at nanomolar concentrations (Sattelle et al., 1980; Harrow et al., 1982, Sattelle et al., 1983). The action of α-bungarotoxin on the cercal afferent input to giant interneurone

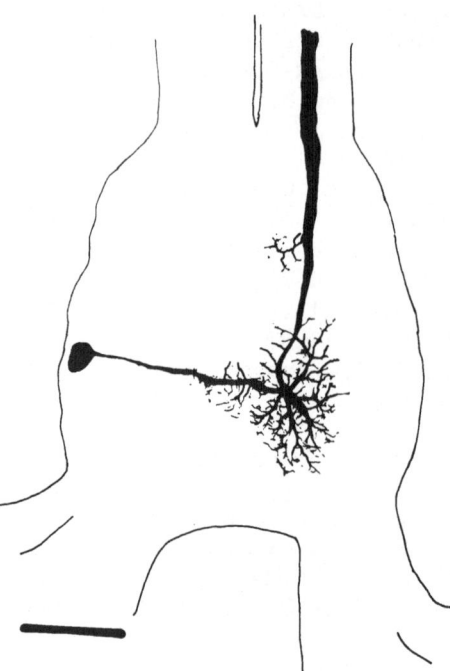

Fig. 1.   Typical morphology of a giant interneurone 2 of the
          cockroach (<u>Periplaneta americana</u>).   Camera lucida
          representation of a silver-intensified, cobalt-backfilled,
          single, giant interneurone in the sixth abdominal ganglion.
          Scale bar represents 200 μm.   Modified from Harrow et al.,
          (1982).

2 (GI 2) has been examined by studying excitatory postsynaptic potentials (EPSPs) recorded by the oil-gap method from GI 2.   When α-
bungarotoxin (1.0×10^{-7} M) was bath perfused over the desheathed
ganglion, the EPSPs recorded in response to the deflection of a long
bristle hair on the lateral-ventral surface of the cercus, progressively reduced in amplitude.   Synaptic transmission was blocked about
90 min after the application of this concentration of α-bungarotoxin.
The compound EPSP recorded in the same experiment in response to
direct electrical stimulation of cercal nerve XI was also completely
blocked by α-bungarotoxin (1.0×10^{-7} M).   After 110 min of exposure
to toxin, increasing the stimulus intensity in order to recruit
further afferents failed to restore synaptic transmission, indicating
that all excitatory afferent input to the giant interneurone via
cercal afferent nerve XI was blocked.   Superfusing the ganglion with
normal saline for 150 min following synaptic block by α-bungarotoxin
did not result in any recovery of synaptic transmission.   In the same
experiment, action potentials elicited by direct stimulation of the
portion of the giant axon in the sixth abdominal ganglion were

unaffected by exposure to 1.0×10^{-7} M α-bungarotoxin. The vertebrate muscarinic antagonist, quinuclidinyl benzilate (QNB) at a concentration of 1.0×10^{-6} M did not affect synaptic transmission between cercal afferents and GI 2, although perfusion of QNB at a concentration of 1.0×10^{-5} M resulted in a partial block of transmission.

Autoradiographic studies of sections of the sixth abdominal ganglion incubated with ^{125}I α-bungarotoxin revealed two separate regions of binding (Fig. 2). In the neuropile region, two distinct patches of specific binding were noted on either side of the midline. Pretreatment with either d-tubocurarine or nicotine at a concentration of 1.0×10^{-3} M (15 min) completely removed these dense patches, indicating that the binding in this area is specific. These regions of neuropile roughly correspond with the location of the cercal afferent, giant interneurone synapses. In addition, dense binding was also observed in the periphery of the ganglion. The distribution of binding in the peripheral regions of the ganglion was not restricted to cell bodies but extended throughout regions occupied by glial cells. Only part of this peripheral binding was removed by pretreatment with 1.0×10^{-3} M d-tubocurarine, thus demonstrating that both specific and non-specific binding components were detected in peripheral regions of the ganglion.

One of the problems of synaptic physiology is that the actions being observed may be the result of an effect on either or both the pre- and postsynaptic terminals. This is also true for the experiments described above, although in this instance it seems likely that α-bungarotoxin is acting postsynaptically as the results can be most simply explained by competitive antagonism of a postsynaptic cholinergic receptor. Insect cell bodies also offer a useful model for studies of receptors. Although they do not possess synapses they do have extrasynaptic receptors which are readily accessible and therefore lend themselves to investigation using ionophoresis and microelectrode techniques. The extrasynaptic cell body receptors of GI 2 have been examined in this way and it has been shown that the depolarizing response to ionophoretically-applied ACh was blocked by micromolar concentrations of α-bungarotoxin (Fig. 3). By contrast micromolar concentrations of QNB were ineffective (Harrow and Sattelle, 1983).

A more accessible cell body than that of GI 2 is that of the fast coxal depressor motorneurone, D_f (Fig. 4). This cell body can be visually located on the ventral surface of the cockroach metathoracic ganglion and has a high sensitivity to ionophoretically-applied ACh (Sattelle et al., 1980; David and Pitman, 1982). The acetylcholine-induced current appears to be carried largely by sodium ions: it is blocked in sodium-free saline (Harrow et al., 1982) and experiments using a sodium-selective microelectrode show a transient increase in intracellular sodium in response to externally-applied ACh (David and Sattelle, unpublished observations).

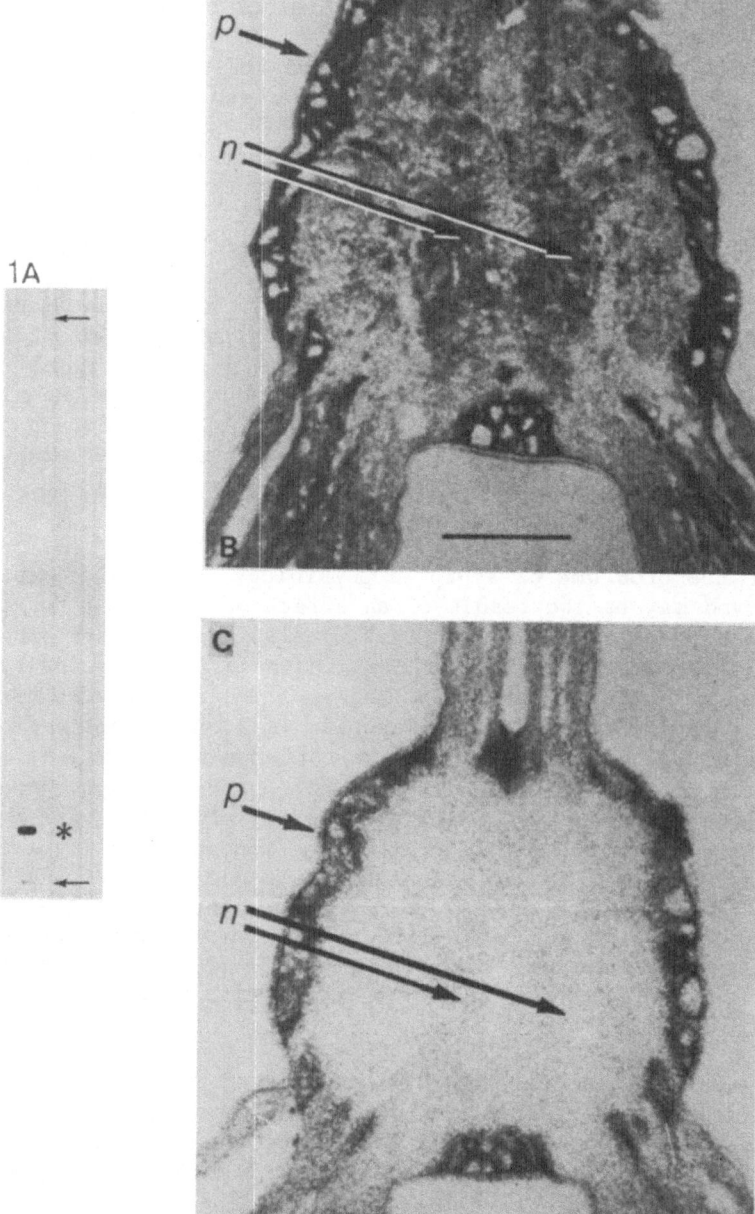

Fig. 2.   (A) SDS-polyacrylamide gel of 20 µg α-bungrotoxin. Top and
          bottom of the gel are indicated by arrows. (B), (C)
          Distribution of ^{125}I α-bungarotoxin in the terminal (sixth)
          abdominal ganglion of an adult male cockroach. Horizontal
          sections of the ganglia are shown and light micrographs were
          prepared with bright-field illumination. (B) In this
          section through the central region of the ganglion the
          silver grains are located primarily in two central patches
          in the neuropile (n) and in the periphery (p) of the
          ganglion. Some of the larger diameter neuronal cell bodies
          can also be seen in the periphery of the section. (C)
          Pretreatment of the sections with 1.0 x 10^{-3} M d-
          tubocurarine chloride for 15 min removed all but the
          background binding of ^{125}I- α-bungarotoxin binding from
          the neuropile (n). Only a portion of the peripheral toxin
          binding was removed (p). Calibration bar = 200 µm (B and
          C). From Sattelle et al., (1983).

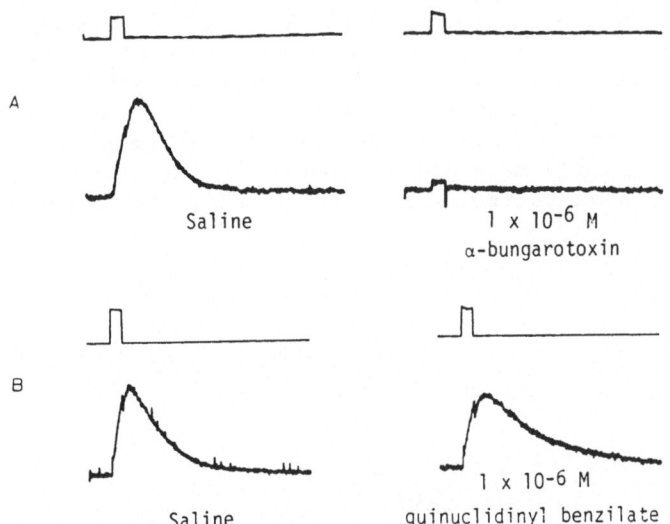

Fig. 3.   Effects of α-bungarotoxin and quinuclidinyl benzilate on the
          response to ionophoretically-applied acetylcholine of the
          cell body membrane of GI 2. (A) 1.0 x 10^{-6} M α-
          bungarotoxin (60 min), ionophoretic dose 25 nC; (B) 1.0 x
          10^{-6} M quinuclidinyl benzilate (60 min), ionophoretic dose
          115 nC. Calibration: (A) vertical (upper) 120 nA, (lower)
          2.5 mV, horizontal 2 s; (B) vertical (upper) 300 nA, (lower)
          2.5 mV, horizontal 2 s. Modified from Harrow and Sattelle
          (1983).

A range of cholinergic antagonists have been tested for their capacity to block the depolarization of the cell body membrane of D_f induced by ionophoretic application of ACh. Nicotinic cholinergic antagonists (α-bungarotoxin, α-cobratoxin, mecamylamine, dihydro-β-erythroidine and benzoquinonium) were particularly effective, while d-tubocurarine and pancuronium were less potent. Hexamethonium, gallamine, decamethonium and succinylcholine were the least effective cholinergic ligands studied, requiring concentrations in excess of 1.0×10^{-3}M to induce a substantial block of the ACh response. The muscarinic ligand, quinuclidinyl benzilate, was only effective on D_f at high concentrations which were substantially different to its dissociation constant as determined in binding studies.

GABA Receptors of Identified Insect Neurones

GABA has been shown to have potent effects on unidentified insect neurones and sub-picomole amounts have been reported to hyper-polarize cockroach neurones (Kerkut et al., 1969; Callec, 1974). This GABA response is thought to be chloride-mediated but there is some evidence that potassium ions may be involved (Kerkut et al., 1969; Hue et al., 1979). Both bicuculline and picrotoxin have been found to block GABA responses (Walker et al., 1971; Hue, 1983), but bicuculline methiodide was found to cause depolarization of unidenti-fied cockroach neurones, while having no effect of the GABA-evoked hyperpolarization (Hue, 1983). Recent studies (Roberts et al., 1981) on unidentified Periplaneta CNS neurones, have shown that dihydro-muscimol is a very effective GABA agonist (14 times GABA activity), muscimol is slightly more potent than GABA, and isoguvacine, isoni-pecotic acid, homomuscimol and thiomuscimol could also mimic the response.

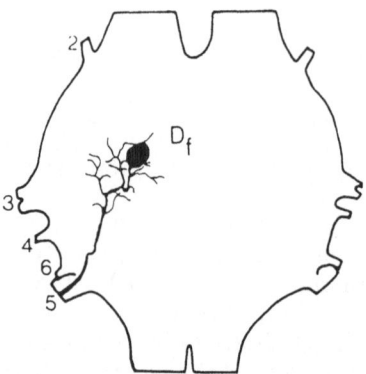

Fig. 4.   Camera-lucida drawing of an HRP-filled fast coxal depressor motorneurone (D_f) in the cockroach metathoracic ganglion. Scale bar indicates 500 μm.   Modified from Sattelle (1985).

Electrophysiological studies have been used to examine GABA responses in the central nervous systems of a wide variety of invertebrates other than the cockroach and various different responses to GABA have been reported. In Aplysia alone five different types of response to GABA have been described (Yarowsky and Carpenter, 1978), and in the suboesophageal ganglionic mass of Helix, differing responses to GABA application were observed on different cells: cells E13 and E16 were excited by GABA whereas cell E4 was inhibited (Walker et al., 1975, 1976; James et al., 1978). These results demonstrate the importance of being able to use identifiable cells when attempting to characterize invertebrate GABA responses.

Studies on the identified motorneurone D_f demonstrated that bath-applied GABA at a concentration of 3.0×10^{-4}M will hyperpolarize the cell. This response could be mimicked by similar concentrations of muscimol, but much higher concentrations of the GABA agonists thiomuscimol and isoguvacine were required to initiate such a response.

Radiolabelled Ligand Binding Characterization of Insect CNS Cholinergic Receptors

[^{3}H] α-Bungarotoxin Binding

The amount of specific [^{3}H] α-BTX binding was linearly proportional to the quantity of CNS extract, in the range 0.05-0.2mg/ml protein, for a given concentration of labelled toxin. Specific binding was saturable (Fig. 5) reaching a maximum at a concentration of approximately 10nM labelled toxin. From Scatchard plots of the data, a dissociation constant (K_d) of 4.8 ± 1.5nM and B_{max} of 910 ± 120 fmoles/mg protein were obtained (Mean values $\pm$ SE (standard error), n=3). Concentrations of labelled toxin up to 50 nM revealed no evidence of further binding sites. A Hill coefficient of 0.99 ± 0.03 (mean $\pm$ SE, n=3) was not significantly different from unity (data from Lummis and Sattelle, 1985).

The specificity of the [^{3}H] α-BTX binding site was assessed by the effectiveness of various cholinergic ligands in preventing binding (Table 1). The most potent inhibitors of binding were unlabelled toxin, d-tubocurarine and nicotine with K_i values of 8.5×10^{-10}M, 8.5×10^{-7}M and 3.5×10^{-6}M respectively, whereas the muscarinic antagonists atropine ($K_i = 2.0 \times 10^{-5}$M) and quinuclidinyl benzilate ($K_i = 1.4 \times 10^{-4}$M) were less effective. (K_i is determined as described in the legend to Table 1). Hexamethonium, an inhibitor of vertebrate ganglionic nicotinic receptors (Volle and Koelle, 1975), and decamethonium, which exerts both agonist and blocking actions at vertebrate neuromuscular nicotinic receptors (Adams and Sakmann, 1978), were almost ineffective on the cockroach preparation ($K_i > 1.0 \times 10^{-4}$M in each case). Table 1 compares the inhibition of [^{3}H] α-BTX binding to various preparations (including Periplaneta

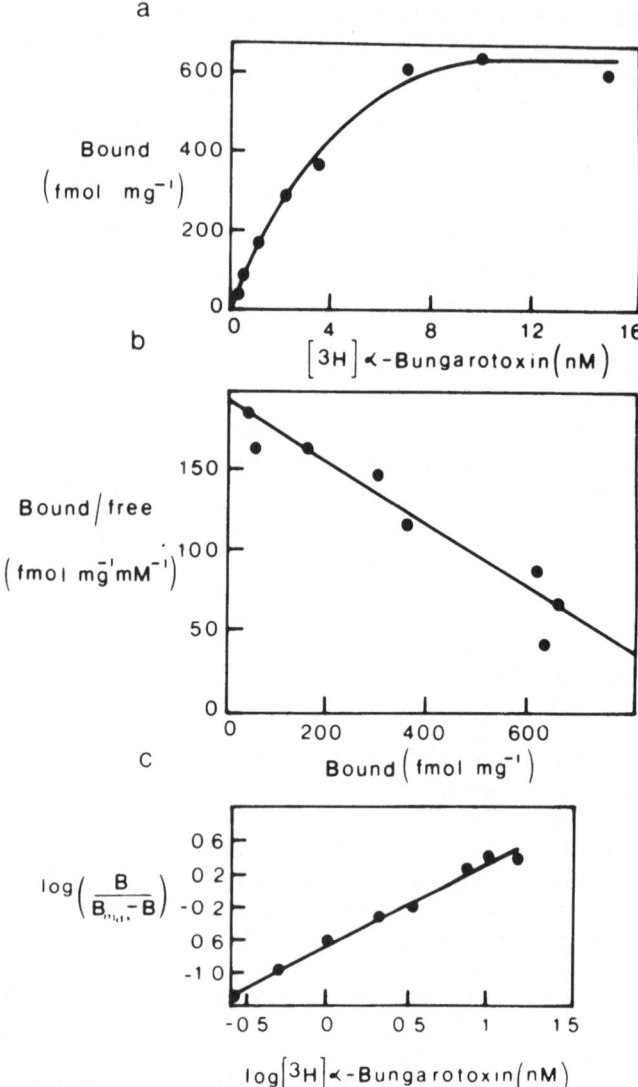

Fig. 5. Binding of [³H] α-BTX ([³H] α-bungarotoxin) to cockroach
nerve cord homogenates. Data from one experiment is shown
but is representative of three similar experiments. (a)
Saturation of specific binding. Nervous tissue homogenates
in buffer were incubated with different concentrations of
radiolabelled α-bungarotoxin for 60 min at 21°C. (b)
Scatchard plot of the specific binding shown in (a). K_d and
B_{max} were determined by least-squares linear regression
analysis of data from three such experiments. (c) Hill plot
of the binding data shown in (a). The Hill coefficient (n_h
was determined using least-squares linear regression analysis. From Lummis and Sattelle (1985).

Table 1. Pharmacology of α-bungarotoxin binding sites

| Ligand | K_i (M) binding | | | | | | V_{50} physiol. |
	optic lobe	rat diaphragm	Aplysia californica (CNS)	Drosophila melanogaster (head)	Locusta migratoria (head ganglia)	Periplaneta americana nerve cords	Periplaneta americana (D_f)
α-bungarotoxin	-	-	-	5×10^{-10}	7×10^{-10}	8.5×10^{-10}	6.4×10^{-6}
nicotine	1.2×10^{-7}	-	-	8×10^{-7}	4×10^{-8}	3.5×10^{-6}	$+2.3 \times 10^{-6}$
d-tubocurarine	2.5×10^{-7}	2.0×10^{-7}	2×10^{-6}	2×10^{-6}	3×10^{-5}	8.5×10^{-7}	8.0×10^{-5}
gallamine	3.8×10^{-6}	-	2×10^{-6}	5×10^{-5}	-	1.4×10^{-5}	1.5×10^{-3}
acetylcholine	3.7×10^{-5}	4.7×10^{-7}	2×10^{-4}	2×10^{-5}	4×10^{-4}	$*1.8 \times 10^{-5}$	$+1.1 \times 10^{-5}$
atropine	8.5×10^{-5}	-	3×10^{-5}	5×10^{-5}	1×10^{-4}	2.0×10^{-5}	1.0×10^{-4}
decamethonium	2.7×10^{-5}	2.1×10^{-6}	2×10^{-4}	-	8×10^{-4}	1.3×10^{-4}	2.8×10^{-3}
hexamethonium	2.5×10^{-4}	1.2×10^{-4}	3×10^{-4}	-	8×10^{-4}	1.4×10^{-4}	8.0×10^{-4}
succinylcholine	-	-	-	-	-	2.8×10^{-4}	2.8×10^{-3}
pancuronium	-	1.1×10^{-7}	-	-	-	2.1×10^{-4}	1.5×10^{-4}
mecamylamine	-	-	-	$>1 \times 10^{-3}$	-	1.0×10^{-3}	2.5×10^{-6}
quinuclidinyl benzilate	-	-	-	-	-	1.4×10^{-4}	1.6×10^{-4}
Reference	Wang et al. (1978)	Colquhoun and Rang (1976)	Shain et al. (1974)	Dudai (1978)	Breer (1981)	Present study	David and Sattelle (1984)

CNS) by cholinergic ligands, and their physiological effectiveness on the fast coxal depressor motorneurone (D_f) in the cockroach meta-thoracic ganglion. In the case of agonists, membrane depolarization is used for comparison, and in the case of antagonists their capacity to block the response to ionophoretically-applied ACh is compared (cf. David and Sattelle, 1984).

[^{3}H]-Quinuclidinyl Benzilate Binding

The specific binding of [^{3}H]-QNB was found to increase linearly with increasing amounts of cockroach nervous tissue up to a concentration of 0.2mg/ml protein. The equilibrium binding curve (Fig. 6) appears to saturate, and concentrations of labelled ligand up to 50nM revealed no evidence of additional binding sites. Scatchard plots of the data yielded a K_d of 8.0 ($\pm$ 3.2) nM and a B_{max} of 138 ($\pm$ 24) fmol/mg (Mean $\pm$ SE, n=3). A Hill coefficient of 0.98 $\pm$ 0.02 was not significantly different from unity.

Ligands active at muscarinic ACh receptors of other species were found to be the most effective in inhibiting binding to cockroach extracts, with dexetimide, atropine and scopolamine being the most potent inhibitors. Nicotinic ligands such as α-bungarotoxin and nicotine were extremely ineffective. The binding site clearly has a muscarinic profile (Table 2), although d-tubocurarine appears to be quite effective in inhibiting [^{3}H]-QNB binding. However it is less potent here than on the toxin binding site and is not as effective as the muscarinic ligands.

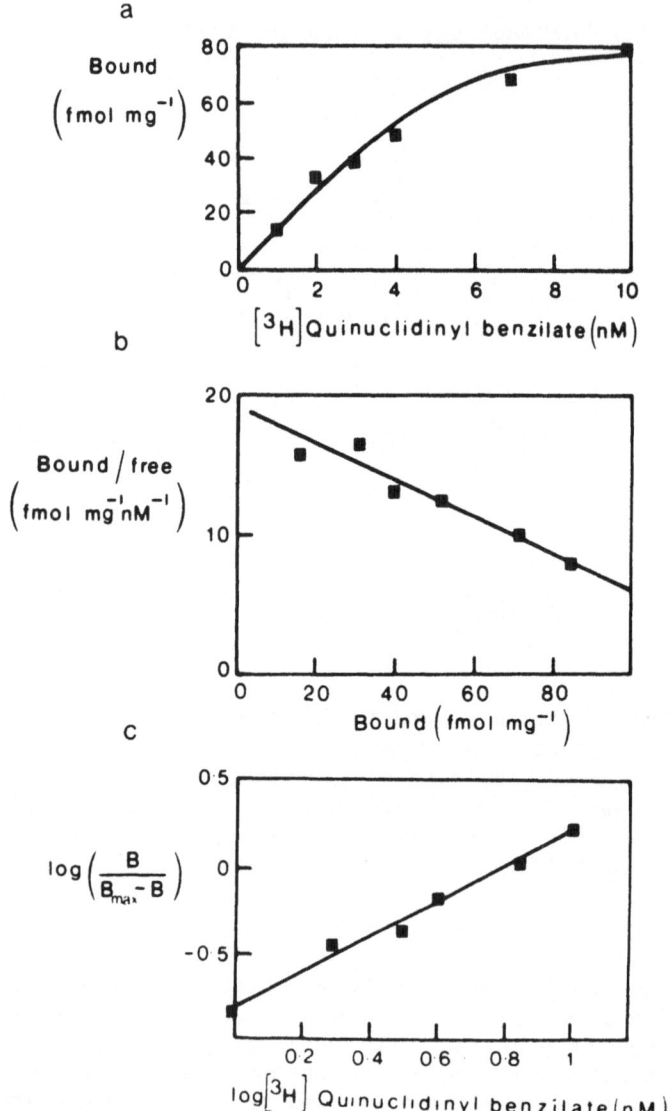

Fig. 6.   Binding of [3H]-QNB ([3H] Quinuclidinyl benzilate) to cockroach nerve cord homogenates. Data from one experiment is
shown but is representative of three similar experiments.
(a) Saturation of specific binding. Nervous tissue homogentes
in buffer were incubated with different concentrations of
radiolabelled quinuclidinyl benzilate for 60 min at 21° C.
(b) Scatchard plot of the specific binding shown in (a). K_d
and B_{max} were determined by least-squares linear regression
analysis of data from three such experiments. (c) Hill plot
of the binding data shown in (a). The Hill coefficient (n_h)
was determined using least-squares linear regression
analysis. From Lummis and Sattelle (1985).

Table 2. Pharmacology of quinuclidinyl benzilate binding sites

Ligand	K_i (M) binding					
	Guinea pig ileum	rat brain	mouse neuroblastoma cells	Drosophila melanogaster (heads)	Logusta migratoria (head ganglia)	Periplaneta americana (nerve cords)
dexetimide	-	-	-	1×10^{-9}	-	5.0×10^{-9}
quinuclidinyl benzilate	2×10^{-10}	-	3.8×10^{-11}	-	-	5.0×10^{-9}
scopolamine	2×10^{-10}	9.5×10^{-10}	4.6×10^{-10}	1×10^{-9}	1×10^{-7}	5.5×10^{-8}
atropine	3×10^{-9}	4.2×10^{-10}	2.2×10^{-10}	4×10^{-7}	4×10^{-7}	1.7×10^{-7}
d-tubocurarine	-	-	-	5×10^{-6}	-	5.0×10^{-6}
carbamylcholine	2×10^{-5}	4.3×10^{-5}	1.2×10^{-5}	3×10^{-5}	1×10^{-4}	2.8×10^{-6}
oxotremorine	6×10^{-7}	8.0×10^{-7}	1.2×10^{-6}	9×10^{-6}	-	1.0×10^{-5}
pilocarpine	6×10^{-7}	-	4.8×10^{-6}	3×10^{-6}	5×10^{-6}	1.1×10^{-5}
α-bungarotoxin	-	-	-	-	-	$>1.0\times10^{-3}$
nicotine	-	-	-	-	-	$>1.0\times10^{-3}$
References	Yamamura and Snyder (1974)	Gilbert et al. (1979)	Strange et al. (1978)	Haim et al. (1978)	Breer (1981)	Present study

Radiolabelled Ligand Binding Studies on Insect CNS GABA Receptors

Radiolabelled ligand binding studies of cholinergic receptors
have been greatly facilitated by the availability of specific, high-
affinity ligand antagonists. To date no such high affinity
antagonists have been discovered for GABA receptors. In insects,
specific radiolabelled GABA binding has been difficult to detect, and
Abalis et al. (1983) have used GABA enhancement of [^{3}H]-flunitrazepam
binding to demonstrate the presence of GABA receptors in housefly
thorax. In vertebrate CNS extensive washing, freeze/thaw, osmotic
shock and/or detergent treatments are used in tissue preparations in
order to reveal GABA receptor sites. These treatments are considered
to remove endogenous inhibitors, including a large proportion of
endogenous GABA. High levels of endogenous GABA have been reported
in insect CNS but treatment of insect CNS tissue with the drastic
procedures described above appeared to destroy GABA binding activity.
In order to preserve this activity and yet remove as much endogenous
GABA as possible, the cockroach CNS crude membrane preparation was
diluted in a vast excess of ice-cold buffer before the final
centrifugation step. Following this procedure, specific GABA binding
was found to increase linearly with increasing tissue concentration
over the range 0.2-20 mg/ml protein. Specific binding of [^{3}H]-GABA
was saturable in the presence of increasing concentrations of GABA
(Fig. 7) and appeared to have a dissociation constant close to 400nM.

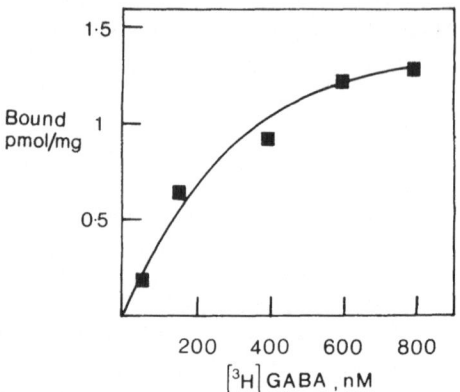

Fig. 7. Binding of [^{3}H]-GABA to cockroach nerve cord homogenates.
Saturation of specific binding. Nervous tissue homogenates
in buffer were incubated with varying concentrations of
radiolabelled GABA for 20 min at 0°C. Non-specific binding
was determined using 1.0 x 10^{-3}M unlabelled GABA or 1.0 x
1.0^{-3}M muscimol. Data from one experiment is shown but is
representative of three similar experiments.

DISCUSSION

Acetylcholine Receptor-Insecticide Interactions

     Several classes of insecticidally-active molecules are active at
ACh receptors (Eldefrawi, 1976; Sattelle, 1980; Corbett et al.,
1984). For example, nicotine one of the earliest insecticides, is a
potent agonist when applied to certain insect neurones (see for
example David and Sattelle, 1984). Yamamoto et al. (1968) showed
that many nicotinoids of 2-pyridylmethylamines were particularly
toxic to houseflies. Nereistoxin (4-N,N dimethylamino-1,2-
dithiolane) isolated from the marine annelid Lumbriconereis
heteropoda is a potent insect neurotoxin (Nitta, 1934; Sakai, 1964;
Narahashi, 1973). Cartap was the first commercial synthetic insecticide based on the structure of a natural toxin not already used as a
pesticide, and in vivo it is metabolized to nereistoxin (Sakai and
Sato, 1971). Nereistoxin blocks cholinergic synaptic transmission
but its action is complex. Eldefrawi et al. (1980) studying Torpedo
electric organ membranes and vertebrate nerve muscle junctions,
concluded that there was an important action on the nicotinic ACh
receptor/ion channel complex but at a site distinct from both the α-
bungarotoxin binding site and the ion channel site. Studies on an
insect nicotinic acetylcholine receptor/ion channel also show that
nereistoxin is extremely active (Sattelle, 1977), though in contrast
to its actions on vertebrate nicotinic sites it is strongly voltage-
dependent in blocking the receptor (Sattelle et al., 1985).

A series of isothiocyanates was synthesized by Baillie et al. (1975) as potential insecticides. These molecules were highly active as choline acetyltransferase inhibitors. When 2-isothiocyanatoethyltrimethylammonium iodide was applied to the isolated abdominal nerve cord of Periplaneta americana, rapid synaptic blocking actions were noted. Synaptic block was the result of postsynaptic depolarization (Sattelle and Callec, 1977), the isothiocyanate under test acting as a cholinergic agonist. For 2-isothiocyanatoethyltrimethylammonium iodide a good correspondence is noted between synaptic blocking action and inhibition of $[^{125}I]$ α-bungarotoxin binding in the CNS of Periplaneta americanca (Gepner et al., 1978). Though this particular molecule of the series is not itself a candidate insecticide because of its charge and water solubility, it seems likely that related compounds of this type may owe part of their toxic actions to an action on nicotinic acetylcholine receptors in addition to their inhibition of the enzyme that synthesizes ACh. Several isothiocyanates and nereistoxin analogues were prepared by Baillie et al., (1978) who showed them to be effective at insect ACh receptors.

A large number of insecticides in current commercial use act by retarding the hydrolysis of ACh by acetylcholinesterase, thereby prolonging the actions of the transmitter at cholinergic synapses. Studies on vertebrate tissues have shown that direct actions on the ACh receptor can be detected for many acetylcholinesterase inhibiting insecticides (cf. Eldefrawi, 1976; Sattelle, 1980). Cholinesterase inhibitors eserine and neostigmine inhibited the binding of radiolabelled α-bungarotoxin to extracts of Drosophila heads (Schmidt-Neilsen et al., 1977); I_{50} values of 1.0×10^{-5} M (eserine) and 2.0×10^{-5} M (neostigmine) were noted. It is possible therefore that certain anticholinesterase insecticides owe some of their toxicity to interactions with ACh receptors.

Pyrethroid insecticides modify the properties of nerve membrane sodium channels (Lund and Narahashi, 1978) and until recently there were no studies on the interactions of these compounds with ACh receptors. Abassy et al. (1983) showed that none of a range of pyrethroids tested significantly inhibited $[^{3}H]$-ACh binding to Torpedo electric organ membranes, but all inhibited $[^{3}H]$-perhydrohistrionicotoxin binding. Ryanodine, a plant alkaloid with insecticidal properties has been shown to affect the rate of nicotinic ACh receptor synthesis by vertebrate skeletal muscle. This action appears to result from changes in internal calcium levels induced by the alkaloid. Finally, chlorobenzilate which is an analogue of DDT and quinuclidinyl benzilate shows some acaricidal effects (see Eldefrawi et al., 1982). It inhibits $[^{3}H]$-quinuclidinyl benzilate binding to housefly head extracts with a K_i of 1.4×10^{-5} M (Shaker and Eldefrawi, 1981) though it is not known if the acaricidal effects of chlorobenzilate are related to its actions on muscarinic cholinergic receptors.

GABA Receptors as a Site of Insecticide Action

    The avermectins are a new chemical class of drugs which have a
novel mode of action against a broad spectrum of nematode and arthropod parasites of animals (Campbell et al., 1983). Ivermectin is the
22,23-dihydro derivative of avermectin $B1_a$ and is in commercial use
in a variety of countries for the treatment and control of parasites
in cattle, horses and sheep.

    Early studies indicated that avermectin $B1_a$ acted neither as a
nicotinic agonist nor as a blocking agent of cholinergic nerve transmission. Further experiments using avermectin in conjunction with
picrotoxin, a GABA agonist, in Ascaris suum suggested that avermectin
$B1_a$ acts by blocking signal transmission from interneurones to
excitatory motoneurones and that GABA is the neurotransmitter that is
blocked (Fritz et al., 1979). This could be due to a stimulation of
GABA-mediated chloride ion conductance by avermectin $B1_a$, as has been
demonstrated in the lobster stretcher muscle, and therefore the
overall effect could be due to (a) avermectin $B1_a$ acting as a GABA
agonist, (b) stimulation of presynaptic GABA release or (c) potentiation of GABA binding to its receptor. Binding studies in mammalian
CNS have shown that avermectin $B1_a$ will stimulate presynaptic GABA
release from rat brain synaptosomes (Pong and Wang, 1980) and will
stimulate GABA binding to rat brain synaptic membranes (Pong et al.,
1982). In order to determine which of these GABA-related effects are
more important to overall antiparasitic activity further experiments,
particularly on the target species, are required.

    The action of pyrethroids, which are widely used insecticides,
have been divided into two groups depending on their symptomology
(Casida et al., 1983). The type I pyrethroid action is similar in
many respects to that of DDT, but the type II snydrome of intra-
cerebrally administered pyrethroids more closely approximates that of
the convulsant picrotoxin. Deltamethrin and a variety of cage convulsants will inhibit the binding of [^{3}H]-dihydropicrotoxin to rat
brain synaptic membranes, suggesting a possible relation between the
type II pyrethroid action and the GABA receptor complex. Using a new
radioligand for the picrotoxin binding site, [^{35}S]-
bicyclophosphorothionate (TBPS), Lawrence and Casida (1983) have
shown that specific binding of this ligand in rat brain synaptic
membranes is inhibited by type II pyrethroids with potencies corresponding to their relative toxicities. They suggest an interaction of
these drugs with the TBPS-picrotoxin receptor although possibly not
at an identical site. In addition to the pyrethroids, three major
classes of polychlorocycloalkane insecticides have been shown to
inhibit [^{35}S]-TBPS binding in a stereospecific and competitive manner
(Lawrence and Casida, 1984). Thus lindane/hexachlorocyclohexane,
toxaphene and aldrin/dieldrin may act at the GABA receptor complex,
possibly as non-competitive GABA antagonists.

The demonstration that a particular insecticide has a high affinity for acetylcholine or GABA receptors does not necessarily indicate that such a membrane protein is the primary site of its toxic action. However in developing a rational approach to the design of insecticides and new leads for synthesis it is instructive to test the actions of both currently-used insecticides and those under development at specific neurotransmitter receptors.

REFERENCES

Abalis, I.M., Eldefrawi, M.E., and Eldefrawi, A.T., 1983, Biochemical identification of putative GABA/Benzodiazepine receptors in house fly thorax muscles, Pestic. Biochem. Physiol., 20:39.

Abassy, M.A., Eldefrawi, M.E., and Eldefrawi, A.T., 1983, Pyrethroid action on the nicotinic acetylcholine receptor/channel, Pestic. Biochem. Physiol., 19:299.

Adams, P.R., and Brown, D.A., 1975, Actions of γ-aminobutyric acid on sympathetic ganglion cells, J. Physiol., 250:85.

Adams, P.R., and Sakmann, B., 1978, Decamethonium both opens and blocks endplate channels, Proc. Nat. Acad. Sci. (U.S.A.), 75:2994.

Alberts, P., and Bartfai, T., 1976, Muscarinic acetylcholine receptor from rat brain: partial purification and characterization, J. Biol. Chem., 251:1543.

Alger, B.E. and Nicoll, R.A., 1980, Epileptiform burst after-hyperpolarizations: calcium dependent potassium potential in hippocampal CA1 pyramidal cells, Science, 210:1122.

Baillie, A.C., Corbett, J.R., Dowsett, J.R., Sattelle, D.B., and Callec, J.J., 1975, Inhibitors of choline acetyltransferase as potential insecticides, Pestic. Sci., 6:645.

Baillie, A.C., Corbett, J.R., and Sharpe, T.M., 1978, The synthesis of potential insecticides designed to bind to the acetylcholine receptor, Pestic. Sci., 9:1.

Barnard, E.A., Beeson, D., Bilbe, G., Brown, D.A., Constanti, A., Conti-Tronconi, B.M., Dolly, D.O., Dunn, S.M.J., Mehraban, F., Richards, B.M. and Smart, T.G., 1983, Acetylcholine and GABA receptors: subunits of central and peripheral receptors and their encoding nucleic acids, Cold Spring Harb. Symp., 48:109.

Barnard, E.A., Dolly, J.O., Lang, B., Lo, M., and Shorr, R.G., 1979, Application of specifically acting toxins to the detection of functional components common to peripheral and central synapses, Adv. Cytopharmacol., 3:409.

Bartfai, T., Berg, P., Schultzberg, M., and Heilbronn, E., 1976, Isolation of a synaptic membrane fraction enriched in cholinergic receptors by controlled phospholipase A_2 hydrolysis of synaptic membranes, Biochim. Biophys. Acta, 426:186.

Birdsall, N.J.M., Burgen, A.S.V., and Hulme, E.C., 1979, A study of the muscarinic receptor by gel electrophoresis, Br. J. Pharmacol., 66:377.

Birdsall, N.J.M., and Hulme, E.C., 1976, Biochemical studies on muscarinic acetylcholine receptors, J. Neurochem., 27:7.

Birdsall, N.J.M., and Hulme, E.C., 1983, Muscarinic receptor sub-classes, Trends Pharm. Sci., 4:451.

Bowery, N.G., Hill, D.R., and Hudson, A.L., 1983, Characteristics of GABA receptor binding sites on rat whole brain synaptic membranes, Br. J. Pharmac., 78:191.

Breer, H., 1981, Properties of putative nicotinic and muscarinic cholinergic receptors in the central nervous system of Locusta migratoria, Neurochem. Int., 3:43.

Callec, J.J., 1974, Synaptic transmission in the central nervous system of insects, in: "Insect Neurobiology," J.E. Treherne, ed., Elsevier North-Holland Biomedical Press, Amsterdam and New York.

Campbell, W.C., Fisher, M.H., Stapley, E.O., Albers-Schonberg, G., and Jacob, T.A., 1983, Ivermectin: a potent new antiparasitic agent, Science, 221:823.

Casida, J.E., Gammon, D.W., Glickman, A.H., and Lawrence, L.J., 1983, Mechanisms of selective action of pyrethroid insecticides, Ann. Rev. Pharmacol. Toxicol., 23:413.

Chiappinelli, V.A., and Zigmond, R.E., 1978, α-Bungarotoxin blocks nicotinic transmission in the avian ciliary ganglion, Proc. Nat. Acad. Sci., (U.S.A.), 75:2999.

Colquhoun, D., and Rang, H.P., 1976, Effects of inhibitors on the binding of iodinated -bungarotoxin to acetylcholine receptors in rat muscle, Mol. Pharmacol., 12:519.

Corbett, J.R., Wright, K., and Baillie, A.C., 1984, "The Biochemical Mode of Action of Pesticides," Academic Press, London.

Curtis, D.R., and Johnston, G.A.R., 1974, Amino acid transmitters in the mammalian central nervous system, Ergebn. Physiol., 69:97.

Curtis, D.R., Duggan, A.W., Felix, D., and Johnston, G.A.R., 1971, Bicuculline, an antagonist of GABA and synaptic inhibition in the spinal cord of the cat, Brain Res., 32:69.

David, J.A., and Pitman, R.M., 1982, The effects of axotomy upon the extrasynaptic acetylcholine sensitivity of an identified motoneurone in the cockroach Periplaneta americana, J. exp. Biol., 98:329.

David, J.A., and Sattelle, D.B., 1984, Actions of cholinergic pharmacological agents on the cell body membrane of the fast coxal depressor motoneurone of the cockroach (Periplanet americana), J. exp. Biol., 108:119.

Dolphin, A.C., 1984, $GABA_B$ receptors: has adenylate cyclase inhibition any functional relevance?, Trends Neuro. Sci., 7:363.

Dudai, Y., 1978, Properties of an α-bungarotoxin-binding cholinergic nicotinc receptor from Drosophila melanogaster, Biochim. Biophys. Acta, 539:505.

Dunlap, K., and Fischbach, G.D., 1981, Neurotransmitters decrease the calcium conductance activated by depolarization of embryonic chick sensory neurones, J. Physiol., 317:519.

Eldefrawi, A.T., 1976, The acetylcholine receptor and its interaction with insecticides, in: "Insecticide Biochemistry and Physiology," C.F. Wilkinson, ed., Plenum Press, New York.

Eldefrawi, A.T., Bakry, N.M., Eldefrawi, M.E., Tsai, M-C., and Albuquerque, E.X., 1980, Neriestoxin interaction with the acetylcholine receptor-ionic channel complex, Mol. Pharmacol., 17:172.

Eldefrawi, A.T., Shaker, N., and Eldefrawi, M.E., 1982, Binding of acetylcholine receptor/ion channel probes to housefly head membranes, in: "Neuropharmacology of Insects," Ciba Foundation Symposium 88, Pitman, London.

Fewtrell, C.M.S., and Rang, H.P., 1973, The labelling of cholinergic receptors in smooth muscle, in: "Drug Receptors," H.P. Rang, ed., Macmillan, London.

Fritz, L.C., Wang, C.C., and Gorio, A., 1979, Avermectin B1$_a$ irreversibly blocks postsynaptic potentials at the lobster neuromuscular junction by reducing muscle membrane resistance, Proc. Nat. Acad. Sci., (U.S.A.), 76:2062.

Gepner, J.I., Hall, L.M., and Sattelle, D.B., 1978, Insect acetylcholine receptors as a site of insecticide action, Nature, 276:188.

Gilbert, R.F.T., Hanley, M.R., and Iversen, L.L., 1979, [^{3}H]-Quinuclidinyl benzilate binding to muscarinic receptors in rat brain: comparison of results from intact brain slices and homogenates, Br. J. Pharmacol., 65:451.

Haga, T., 1980, Molecular size of muscarinic acetylcholine receptors of rat brain, FEBS Lett., 113:68.

Haga, T., 1983, Characterization of muscarinic acetylcholine receptors solubilized by L-γ-Lysophosphatidylcholine and Lubrol PX, in: "Pharmacological and Biochemical Aspects of Neurotransmitter Receptors," H. Yoshida and H.I. Yamamura, eds., John Wiley & Sons, U.K.

Haim, N., Nahum, S., and Dudai, Y., 1979. Properties of a putative muscarinic cholinergic receptor from Drosophila melanogaster, J. Neurochem., 32:543.

Harrow, I.D., David, J.A., and Sattelle, D.B., 1982, Acetylcholine receptors of identified insect neurones, in: "Neuropharmacology of Insects," Ciba Foundation Symposium 88, Pitman, London.

Harrow, I.D., Hue, B., Gepner, J.I., Hall, L.M., and Sattelle, D.B., 1980, An α-bungarotoxin sensitive acetylcholine receptor in the CNS of the cockroach, Periplaneta americana L., in: "Insect Neurobiology and Pesticide Action," Proc. Soc. Chem. Ind.

Harrow, I.D., and Sattelle, D.B., 1983, Acetylcholine receptors on the cell body membrane of giant interneurone 2 in the cockroach Periplaneta americana, J. Exp. Biol., 105:339.

Hue, B., 1983, Quelques aspects electrophysiologiques et pharmacologiques de l'inhibition postsynaptique dans le systeme nerveux central de la blatte, publication 10, Ph.D. Thesis., L'Université d'Angers, La France.

Hue, B., Pelhate, M., and Chanelet, J., 1979, Pre and postsynaptic effects of taurine and GABA in the cockroach central nervous system, Can. J. Neurol. Sci., 6:243.

James, V.A., Krogsgaard-Larsen, P., and Walker, R.J., 1978, The action of conformationally restricted analogues of GABA on Limulus and Helix central neurones, Experientia, 34:1630.

Kass, I.S., Wang, C.C., Walrond, J.P., and Stretton, A.O.W., 1979, Avermectin Bl$_a$, a paralyzing anthelmintic that affects inter-neurones and inhibitory motoneurones in Ascaris., Proc. Nat. Acad. Sci., (U.S.A.), 77:6211.

Kerkut, G.A., Pitman, R.M., and Walker, R.J., 1969, Iontophoretic application of acetylcholine and GABA onto insect central neurones, Comp. Biochem. Physiol., 31:611.

Kloog, Y., and Sokolowsky, M., 1977, Muscarinic acetylcholine receptor interactions: competition binding studies with agonist and antagonists, Brain Res., 134:167.

Kloog, Y., and Sokolowsky, M., 1978, Studies on muscarinic acetylcholine receptors from mouse brain: characterization of the interaction with antagonists, Brain Res., 144:31.

Krnjevic, K., 1974, Chemical nature of synaptic transmission in vertebrates, Physiol. Rev., 54:418.

Lawrence, L.J., and Casida, J.E., 1983, Stereospecific action of pyrethroid insecticides on the γ-aminobutyric acid receptor-ionophore complex, Science, 221:1399.

Lawrence, L.J., and Casida, J.E., 1984, Interactions of Lindane, Toxaphene and cyclodienes with brain-specific t-butylbicyclophosphorothionate receptor, Life Sci., 35:171.

Lummis, S.C.R., and Sattelle, D.B., 1985, Binding of N-[propionyl-^{3}H]-propionylated α-bungarotoxin and L-[benzilic-4,4'-^{3}H]-quinuclidinyl benzilate to CNS extracts of the cockroach Periplaneta americana, Comp. Biochem. Physiol., 80C:75.

Lund, A.E., and Narahashi, T., 1981, Interaction of DDT with sodium channels in squid giant axon membranes, Neurosci., 6:2253.

Morley, B.J., Kemp, G.E., and Salvaterra, P., 1979, α-bungarotoxin binding sites in the CNS, Life Sci., 24:859.

Narahashi, T., 1973, Mode of action of nereistoxin on excitable tissues, in: "Marine Pharmacognosy Actions of Marine Biotoxins at the Cellular Level," Academic Press, New York.

Nathanson, N.M., 1982, Regulation and development of muscarinic acetylcholine receptors, Trends Neuro. Sci., 5:401.

Newberry, N.R., and Nicoll, R.A., 1984a, Direct hyperpolarizing action of baclofen on hippocampal pyramidal cells, Nature, 308:450.

Newberry, N.R., and Nicoll, R.A., 1984b, A bicuculline-resistant inhibitory post-synaptic potential in rat hippocampal pyramidal cells in vitro, J. Physiol., 348:239.

Nitta, S., 1934, Uber Nereistoxin einen giften Bestandteil von Lumbriconereis heteropoda Marenz (Eunicidae), Yakaguku Zasshi, 54:648.

Olsen, R.W., 1982,   Drug interactions at the GABA receptor-ionophore complex, Ann. Rev. Pharmacol. Toxicol., 22:245.

Oswald, R.E. and Freeman, J.A., 1981,   Alpha-bungarotoxin binding and central nervous system nicotinic acetylcholine receptors, Neurosci., 6:1.

Pong, S.-S., Dehaven, R., and Wang, C.C., 1981,   Stimulation of benzodiazepine binding to rat brain membranes and solubilized receptor complex by avermectin Bl_a and γ-aminobutyric acid, Biochim. Biophys. Acta., 646:143.

Pong, S.-S., and Wang, C.C., 1980,   The specificity of high affinity binding of avermectin Bl_a to mammalian brain, Neuropharmacol., 19:311.

Popot, J.-L., and Changeux, J.-P., 1984,   Nicotinic receptor of acetylcholine: structure of an oligomeric integral membrane protein, Physiol. Rev., 64:1162.

Roberts, C.J., Krogsgaard-Larsen, P., and Walker, R.J., 1981, Studies on the action of GABA, muscimol and related compounds on Periplaneta and Limulus neurones, Comp. Biochem. Physiol., 69C:7.

Roberts, E., Chase, T.N., and Tower, D.B., 1976, in: "GABA in Nervous System Function," Raven Press, New York.

Sakai, M., 1964,   Studies on the insecticidal action of nereistoxin, 4-N,N-dimethylamino-1,2-dithiolane.   I. Insecticidal properties, Jap. J. Appl. Ent. Zool., 8:324.

Sakai, M., and Sato, Y., 1971,   Metabolic conversion of the nereistoxin-related compounds into nereistoxin as a factor of their insecticidal action, in: "Abst. 2nd Int. Cong. Pestic. Chem.," Tel Aviv.

Salvaterra, P.M. and Foders, R.M., 1979, [^{125}I] α-bungarotoxin binding and [^{3}H]-quinuclidinyl benzilate binding in central nervous systems of different species, J. Neurochem., 32:1509.

Sattelle, D.B., 1977,   A simple assay for the actions of toxic agents on synaptic transmission in the insect C.N.S., in "Crop Protection Agents - Their Biological Evalution", N.R. McFarlane, ed., Academic Press, London.

Sattelle, D.B., 1980,   Acetylcholine receptors of insects, Adv. Insect. Physiol., 15:215.

Sattelle, D.B., and Callec, J.J., 1977,   Actions of isothiocyanates on the central nervous system of Periplaneta americana, Pestic. Sci., 8:735.

Sattelle, D.B., David, J.A., Harrow, I.D., and Hue, B., 1980, Actions of α-bungarotoxin on identified insect central neurones, in: "Receptors for Neurotransmitters, Hormones and Pheromones in Insects," D.B. Sattelle, L.M. Hall, and J.G. Hildebrand, eds., Elsevier North-Holland Biomedical Press, Amsterdam.

Sattelle, D.B., Harrow, I.D., David, J.A., Pelhate, M., Callec, J.J., Gepner, J.I., and Hall, L.M., 1985, Nereistoxin: actions on a CNS acetylcholine receptor/ion channel in the cockroach Periplaneta americana, J. exp. Biol., in press.

Sattelle, D.B., Harrow, I.D., Hue, B., Pelhate, M., Gepner, J.I., and
    Hall, L.M., 1983, α-Bungarotoxin blocks excitatory synaptic
    transmission between cercal sensory neurones and giant
    interneurone 2 of the cockroach Periplaneta americana, J. exp.
    Biol., 107:473.

Schmidt-Nielsen, B.K., Gepner, J.I., Teng, N.N.H., and Hall, L.M.,
    1977, Characterization of an α-bungarotoxin binding component
    from Drosophila melanogaster, J. Neurochem., 29:1013.

Shain, W., Greene, L.A., Carpenter, D.O., Sytkowski, A.J., and Vogel,
    Z., 1984, Aplysia acetylcholine receptors: blockade by and
    binding of α-bungarotoxin, Brain Res., 72:225.

Shaker, N., and Eldefrawi, A.T., 1981, Muscarinic receptor in house
    fly brain and its interaction with chlorobenzilate, Pestic.
    Biochem. Physiol., 15:14.

Sokolowsky, M., Gurwitz, D., and Kloog, J., 1983, Biochemical
    characterization of the muscarinic receptors, Adv. Enzymol.,
    55:137.

Strange, P.G., Birdsall, N.J.M., and Burgen, A.S.V., 1978, Ligand-
    binding properties of muscarinic acetylcholine receptors in
    mouse neuroblastoma cells, Biochem. J., 172:495.

Volle, R.E., and Koelle, G.B., 1975, Ganglionic stimulating and
    blocking agents, in: "The Pharmacological Basis of
    Therapeutics," L.S. Goodman and A. Gilman, eds., Macmillan,
    London.

Wang, G.K., Molinaro, S., and Schmidt, J., 1978, Ligand responses of
    α-bungarotoxin binding sites from skeletal muscle and optic lobe
    of the chick, J. Biol. Chem., 253:8507.

Yamamoto, I., Soeda, Y., Kamimura, H., and Yamamoto, R., 1968,
    Studies on nicotinoids as an insecticide. VII. Cholinersterase
    inhibition by nicotinioids and pyridylalkylamines - its
    significance to mode of action, Agr. Biol. Chem., 32:1341.

Yamamura, H.I., and Snyder, S.H., 1974, Muscarinic cholinergic
    receptor binding in rat brain, Proc. Nat. Acad. Sci. (U.S.A.),
    71:1725.

Yamamura, H.I., Enna, S.J., and Kuhar, M.J., 1978, "Neurotransmitter
    Receptor Binding," Raven Press, New York.

Yarowsky, P.J., and Carpenter, D.O., 1978, Receptors for gamma-
    aminobutyric acid (GABA) on Aplysia neurones, Brain Res.,
    144:75.

BRIDGED BICYCLIC ORGANOPHOSPHORUS COMPOUNDS AS A PROBE FOR

TOXICOLOGICAL STUDY ON GABA SYNAPSE

Yoshihisa Ozoe and Morifusa Eto

Department of Environmental Sciences
Shimane University, Matsue 690, Japan
Department of Agricultural Chemistry
Kyushu University, Fukuoka 812, Japan

INTRODUCTION

Bridged bicyclic organophosphorus compounds refer to the analogs of 2,6,7-trioxa-1-phosphabicyclo[2.2.2]octane in this chapter. The bicyclic compounds include phosphites (I: X = P), phosphates (I: X = P=O) and phosphorothionates (I: X = P=S) (Fig. 1). A symmetrical "cage" structure with high dipole moment is characteristic of those compounds. The bicyclic phosphites are used in the field of coordination chemistry (Verkade, 1972, 1973) and some of them (e.g., the 4-ethyl analog) are commercially available. In 1970, unexpectedly high toxicity of the 4-ethyl bicyclic phosphite (I: R = C_2H_5, X = P) was noticed by Gage (1970) who investigated the inhalation toxicity of 109 industrial chemicals. Bellet and Casida (1973) were the first to report that the toxicity of the bicyclic phosphorus esters is not due to the inhibition of acetylcholinesterase unlike organophosphorus insecticides. This fact makes those esters quite unique. Later, the 4-ethyl bicyclic phosphate (I: R = C_2H_5, X = P=O) was detected by Petajan et al. (1975) and Voorhees et al. (1975) as a toxic principle produced in smoke when a fire-retarded polyurethane foam was combusted.

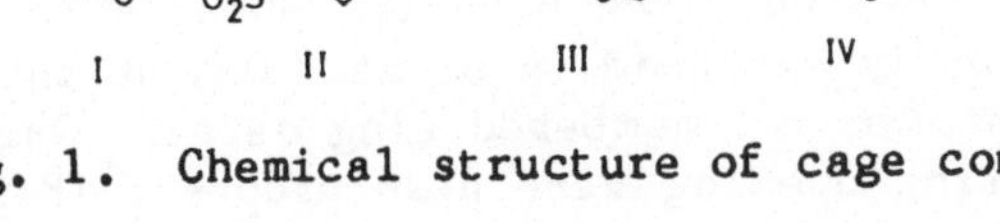

Fig. 1. Chemical structure of cage compounds

The authors' interest was focused on the characteristic structure and unknown mode of action of the cage compounds. Thus, first of all, some bicyclic phosphate analogs were synthesized to do the following experiments (Ozoe and Eto, 1982a). Biological activity of organophosphorus compounds is in many cases related to their phosphorylating or alkylating ability (Eto, 1974). The rate and mechanism of hydrolysis, methanolysis and serine enzyme inhibition were examined to evaluate the reactivity of the bicyclic phosphates (BPs) (Ozoe et al., 1982b, 1982d). Structure-toxicity relationship of bridgehead-substituted BPs was analyzed quantitatively by Hansch-Fujita's regression analysis and mammalian toxicity was compared with toxicity to insects (Ozoe et al., 1983). Since the symptoms of BP poisoning indicated the involvement of the nervous system in the toxic action, the effect of BPs on the invertebrate and vertebrate neuro-muscular junction was investigated (Korenaga et al., 1977). The binding of BPs to a sort of receptor seemed to be important in exerting toxicity. Thus a ligand-receptor binding assay was done by use of synaptic membrane fraction of the rat brain (Ozoe et al., 1982c). An interaction of BPs with a critical target site was speculated by use of macrocyclic compounds as a simple BP binding site model (Ozoe et al., 1981). This chapter summarizes our findings obtained from several aspects on the bicyclic phosphates and intends to gain a better understanding of the mode of action from recent works. Our investigations were confined to the bicyclic phosphates because there was no essential difference in toxicity among the bicyclic phosphites, phosphates and phosphorothionates with the same substituents (Bellet and Casida, 1973).

REVIEW OF RELATED WORK

Toxicity of BPs drastically changes with modification of the bridgehead substituent. One can see the variation in the data reported by Bellet and Casida (1973). Quantitative analysis of structure-toxicity relationship is described in the following section of this chapter. For instance, 4-isopropyl BP was 33 times as toxic as the organophosphorus pesticide parathion while 4-methyl BP was 5 times less toxic than parathion and 4-hydroxymethyl BP nontoxic. The remarkable dependence of toxicity on the bridgehead substituent is in contrast to the effect of the phosphoryl moiety. Distribution and metabolic fate of a toxic BP were compared with those of a less toxic BP from the point of view that differences in toxicity between BP analogs could be explained by difference in localization and detoxification. For metabolic study the paper of Milbrath et al. (1978) should be referred to. The bicyclic phosphates and metabolites were rapidly excreted in urine. One of the metabolites identified was a nontoxic six-membered ring ester. Oxidation and further hydrolytic ring opening were also proposed (Fig. 2). An extremely toxic analog (4-t-butyl BP) was converted to nontoxic hydrolysis products to a greater extent than an alalog of low

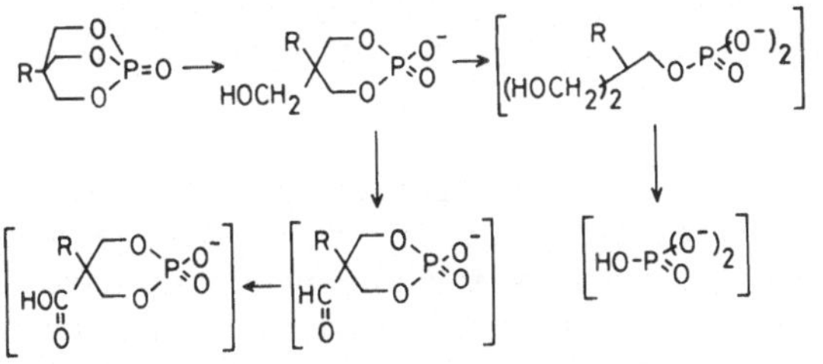

Fig. 2.   Biotransformation pathway of bicyclic phosphates
          (Milbrath et al., 1978).

toxicity (4-methyl BP). The level of 4-t-butyl BP in the mouse brain
was only two to five times as high as that of 4-methyl BP.
Approximately 180-fold higher toxicity of the former compound than of
the latter can not be explained by difference in localization and
detoxification. Difference in the mode of interaction or in affinity
of BPs with a target site may contribute to difference in toxicity.

Another point to be mentioned here is the toxicity of related
cage compounds (Casida et al., 1976; Milbrath et al., 1979). These
compounds are bicyclic orthocarboxylic acid esters (analogs of 2,6,7-
trioxabicyclo[2.2.2]octane (I: X = CH), tetramethylenedisulphotetramine (II, 2,6-dithia-1,3,5,7-tetrazatricyclo [3.3.1.1^{3,7}]decane
2,2,6,6-tetraoxide) and p-chlorophenyl silatrane (III,1-(p-
chlorophenyl)-2,8,9-trioxa-5-aza-1-silabicyclo[3.3.3]undecane). The
toxicity of the bicyclic orthocarboxylates was dependent on both the
C-1 and C-4 substituents. The most toxic compound was the 1-phenyl-
4-t-butyl analog (I: R = t-C_4H_9, X = C-C_6H_5; mouse ip LD$_{50}$, 1.3
mg/kg). It is interesting to note that there were two optimum sizes
as to the C-1 substituent. Relatively high toxicity was conferred by
small (eg., H and CH_3) and large (eg., C_4H_9 and C_6H_5) C-1
substituent. Bicyclic orthocarboxylates with medium C-1 substituent
had little toxicity. This finding is discussed in connection with
the mode of action in the following section. Mouse intraperitoneal
LD$_{50}$ values of tetramethylenedisulphotetramine and p-chlorophenyl
silatrane were reported to be 0.24 mg/kg and 0.22 mg/kg, respectively
(Casida et al., 1976). These compounds produced symptoms of
poisoning similar to those of the bicyclic organophosphorus esters
and inhibited the binding of the radioactive analog of the γ-
aminobutyric acid (GABA) antagonist picrotoxinin to specific sites in
the rat and American cockroach brains, as well as BPs (Ticku and
Olsen, 1979; Tanaka et al., 1984). Tetramethylenedisulphotetramine
was shown to antagonize the action of GABA in the rat superior
cervical ganglion and the crab neuromuscular junction (Bowery et al.,
1975; Large, 1975).

EXPERIMENTAL

Synthesis

As mentioned in the preceding section, the toxicity of BPs is greatly dependent on the size and nature of the bridgehead substituent whereas no essential difference is observed in toxicity among the phosphites, phosphates and phosphorothionates with the same bridgehead substituent. Minor modification of the bridgehead substituent results in a drastic variation in toxicity. To understand the structure-toxicity relationship quantitatively, BPs with various substituents were synthesized (Ozoe et al., 1982a).

Most of BPs were able to be synthesized by reaction of the corresponding triol with phosphoryl chloride in the presence of pyridine in acetonitrile (direct method).

$$RC(CH_2OH)_3 \quad + \quad POCl_3 \quad \longrightarrow \quad RC(CH_2O)_3P=O$$

The Tollens condensation, i.e., the mixed aldol condensation followed by the Cannizzaro reaction, is a well known method for the synthesis of triols, but an alternative method was needed to prepare PBs with a variety of substituents.

$$RCH_2CHO \quad + \quad 3\ HCHO \quad \longrightarrow \quad RC(CH_2OH)_3$$

Alkylmalonates served as starting materials for the alternative method. Base-catalyzed hydroxymethylation of alkylmalonates with paraformaldehyde in dimethyl sulfoxide followed by lithium aluminum hydride reduction of the resultant hydroxymethylated malonates gave triols with branched alkyl groups.

$$RCH(COOR'')_2 \quad + \quad (CH_2O)_n \longrightarrow RC(CH_2OH)(COOR'')_2 \longrightarrow RC(CH_2OH)_3$$

The above two methods give triols for 4-substituted BPs. Triols for 3-methyl-4-alkyl BPs were prepared by application of a similar method to α-alkylacetoacetates.

$$CH_3COCH(R')COOR'' \longrightarrow CH_3COC(R')(CH_2OH)COOR'' \longrightarrow R'C(CH(OH)CH_3)(CH_2OH)_2$$

The acylation of diethyl malonate with pivaloyl chloride gave diethyl pivaloylmalonate, which was reduced with lithium aluminum hydride to give the triol for 3-t-butyl BP.

$$(CH_3)_3CCOCH(COOR'')_2 \quad \longrightarrow \quad (CH_3)_3CCH(OH)CH(CH_2OH)_2$$

Similar methods are applicable to so-called active methylene compounds to give a variety of triol intermediates for the synthesis of BPs.

Some BPs can be prepared by modification of the substituents of BP analogs, although BPs occasionally undergo ring-opening on chemical treatment, especially under alkaline conditions. For instance, 4-acetyl BP was allowed to react with sodium borohydride and methylmagnesium bromide to afford 4-(1-hydroxyethyl) and 4-(1-hydroxy-1-methylethyl) analogs, respectively (Casida et al., 1976; Ozoe et al., 1982a).

$$CH_3COC(CH_2O)_3P{=}O \nearrow \begin{matrix} CH_3CH(OH)C(CH_2O)_3P{=}O \\ \\ (CH_3)_2(OH)CC(CH_2O)_3P{=}O \end{matrix}$$

The indirect method is sometimes useful for synthesis of PBs which can not be prepared by the direct method. Table 1 lists representative PBs synthesized.

Chemical Reactivity

There are many toxicants or pesticides whose potencies are associated with their chemical reactivity. This is the case for biologically active organophosphorus compounds. The activity of organophosphorus insecticides is due to inhibition of acetylcholinesterase. As for phosphorus esters which exhibit delayed neurotoxicity, inhibition of the "neurotoxic esterase" is involved in the primary action (Eto, 1974). The inhibition is the results of phosphorylation of the active site of the enzymes. It was noticed in the course of the synthesis of BPs that they were relatively labile to alkali whereas resistant to acid. This finding suggested the possibility that phosphorylation of a critical target site may be involved in the mode of action, and led us to investigate the reactivity of BPs.

The bicyclic phosphates easily underwent two-step methanolysis (Ozoe et al., 1982b).

$$CH_3{-}C(CH_2O)_3P{=}O \longrightarrow HOCH_2(CH_3)C(CH_2O)_2P(O)OCH_3 \longrightarrow (HOCH_2)_2(CH_3)CCH_2OP(O)(OCH_3)_2$$

The pseudo-first-order rate constants for the methanolysis of 4-methyl BP and the resultant monocyclic ester were estimated to be 1.16×10^{-1} min^{-1} and 9.80×10^{-3} min^{-1} respectively at a concentration of 0.125 M sodium hydroxide in aqueous methanol at $31^\circ C$.

The bicyclic phosphates were readily hydrolyzed to monocyclic partial esters (Ozoe et al., 1982b).

$$C_2H_5{-}C(CH_2O)_3P{=}O \longrightarrow HOCH_2(C_2H_5)C(CH_2O)_2P(O)OH$$

Table 2 shows the pseudo-first-order rate constants for the hydrolysis of BPs at $25^\circ C$ in 0.03 M sodium hydroxide solution. The

Table 1. Representative bicyclic phosphates synthesized

R^1	R^2	Mp(°C)	R^1	R^2	Mp(°C)
H	H	226–228	C_6H_5	H	246
CH_3	H	245	OCH_3	H	157
C_2H_5	H	203	CH_2OCH_3	H	136.5–137.5
C_3H_7	H	210–211	$CH(OCH_3)_2$	H	155–156
$i-C_3H_7$	H	240	CH_2Br	H	147–149
C_4H_9	H	196–197	CH_2Cl	H	170–171.5
$i-C_4H_9$	H	165–168	CH_3	CH_3	152–154
$s-C_4H_9$	H	171–173	C_2H_5	CH_3	126–127
$t-C_4H_9$	H	>245	C_3H_7	CH_3	128–129
C_5H_{11}	H	137.5–138	H	$t-C_4H_9$	165
$i-C_5H_{11}$	H	134–136	$N(CH_3)_2$	H	210 (decomp.)
$CH(CH_3)C_3H_7$	H	83–84	$COCH_3$	H	187–188
$CH(C_2H_5)_2$	H	122–123	NO_2	H	241–242
$c-C_6H_{11}$	H	228–229	$CH(OH)CH_3$	H	179–180
$CH_2CH=CH_2$	H	167–169			

value of 4-t-butyl BP is comparable to that of fenitroxon.

The phosphorylation of acetylcholinesterase by organophosphorus compounds is caused by a mechanism similar to alkaline methanolysis and hydrolysis. Since BPs were highly reactive in an alkaline solution, they were expected to exhibit high inhibitory activity against acetylcholinesterase. However, anticholinesterase data revealed that BPs were very poor inhibitors (Table 3) (Ozoe et al., 1982d). It was supposed that low inhibitory activity was attributable to low affinity of the compounds to the enzyme. Mechanism of inhibition was analyzed by the method of Hart and O'Brien (1973) to evaluate phosphorylating ability and affinity to the enzyme separately. Table 4 shows the inhibition constants (dissociation constant K_d, phosphorylation constant k_2 and bimolecular reaction constant k_i) of 4-nitro BP, indicating that the poor inhibitory activity is ascribed to poor phosphorylating ability as well as low affinity. The bicyclic phosphate was approximately

Table 2. First-order hydrolysis rate constants of bicyclic phosphorus esters and related compounds at $25^{\circ}C$ in 0.03 M sodium hydroxide solution

$R-C(CH_2O)_3P=X$		Rate constants (min^{-1})
R	X	
C_2H_5	O	3.35×10^{-2}
$n-C_4H_9$	O	2.48×10^{-2}
$t-C_4H_9$	O	4.29×10^{-2}
C_2H_5	S	4.14×10^{-3}
Fenitroxon		4.23×10^{-2} [a]
Fenitrothion		3.90×10^{-3} [b]

a Evaluated from the second-order rate constant 1.41 $min^{-1}M^{-1}$ (Mochida et al., 1976).
b Evaluated from the second-order rate constant 1.30×10^{-1} $min^{-1}M^{-1}$ (Mochida et al., 1976).

Table 3. Anticholinesterase activity of selected bicyclic phosphates (1.5 mM) against bovine erythrocyte (B-AChE) and housefly head (H-AChE) acetylcholinesterases

R^1	R^2	Inhibition (%)	
		B-AChE	H-AChE
CH_3	H	0	2
$n-C_3H_7$	H	3	8
$i-C_3H_7$	H	6	8
$t-C_4H_9$	H	6	16
$CH(C_2H_5)_2$	H	9	32
$c-C_6H_{11}$	H	12	19
C_6H_5	H	10	14
NO_2	H	37	38
$n-C_3H_7$	CH_3	2	25
H	$t-C_4H_9$	13	26

Table 4.   Kinetic constants of 4-nitro BP and fenitroxon for the
           inhibition of bovine erythrocyte (B-AChE) and housefly
           head (H-AChE) acetylcholinesterases

constants	B-AChE	H-AChE
	4-NO$_2$-BP	
K_d (M)	$(1.88 \pm 0.23) \times 10^{-2}$ [a]	$(5.28 \pm 0.83) \times 10^{-3}$ [c]
k_2 (min^{-1})	2.40×10^{-2} [b]	$(3.43 \pm 0.55) \times 10^{-2}$ [c]
k_i (min^{-1}M^{-1})	1.60 [b]	7.61 ± 1.30 [d]
	Fenitroxon [e]	
K_d (M)	6.7×10^{-5}	1.1×10^{-5}
k_2 (min^{-1})	5.0	8.3
k_i (min^{-1}M^{-1})	0.73×10^5	7.6×10^5

a The value is the mean ± S.E. of four determinations.
b The value could be determined only at a high concentration of
  4-nitro BP (10 mM).
c Each value is the mean ± S.E. of data from three separate
  experiments, each performed in quadruplicate.
d The value is the mean ± S.E. of k_i value calculated from each
  K_d and k_2.
e Hollingworth et al. (1967).

10^2-10^3 times less active in phosphorylating acetylcholinesterase
than other usual organophosphorus inhibitors such as paraoxon. The
bicyclic phosphates were also very poor inhibitors against α-
chymotrypsin.

Structure-Toxicity Relationship

    The bicyclic phosphates yielded convulsive seizures and death
in animals. Table 5 lists LD$_{50}$ (median lethal dose) values of
BPs against mice and houseflies. In 1976, we found the extremely
high toxicity of 4-t-butyl BP to mice (Eto et al., 1976). It was
about 3.4-fold more toxic than 4-isopropyl BP which had been reported
as the most toxic analog. The bicyclic phosphate elicited the most
potent convulsant activity among analogs (Bowery et al., 1977). The
dependence of toxicity on the bridgehead substituent was evident.
The bulkiness or branching of the bridgehead substituent seemed to be
important for high toxicity. The bicyclic phosphates with hydrophilic
bridgehead substituents were of low toxicity. Hansch-Fujita's
regression analysis (Hansch and Fujita, 1964) was applied to
representative bridgehead-substituted BPs (Table 6) and the
equation on page 84 was derived as the most refined one for mammalian
(mouse) toxicity (Ozoe et al., 1983).

Table 5. Toxicity of bicyclic phosphates applied by injection to houseflies (A and B) and mice (C)

R^1	R^2	LD_{50} (μg/fly)[a] Alone (A)	LD_{50} (μg/fly)[a] With pb (B)	SR[b] (A/B)	LD_{50} (mg/kg)[c] (C)
CH_3	H	>0.75	25.1	–	32
C_2H_5	H	>0.75	0.913	–	1.0
C_3H_7	H	0.449	0.169	2.66	0.38
$i\text{-}C_3H_7$	H	>1.0	0.186	–	0.18
C_4H_9	H	>1.5	0.360	–	1.5
$i\text{-}C_4H_9$	H	5.603	0.168	33.35	0.24
$s\text{-}C_4H_9$	H	1.575	0.145	10.86	0.21
$t\text{-}C_4H_9$	H	0.705	0.124	5.69	0.053
C_5H_{11}	H	>0.75	NT[e]	–	37
$i\text{-}C_5H_{11}$	H	>0.75	NT	–	11.25
$CH(CH_3)C_3H_7$	H	>0.75	0.929	–	4.4
$CH(C_2H_5)_2$	H	NT	0.505	–	0.80
$COCH_3$	H	>0.75	NT	–	51
$CH(OH)CH_3$	H	NT	>40.0	–	112.5
OCH_3	H	>0.75	9.760	–	8.0
$CH(OCH_3)_2$	H	>0.75	NT	–	70
CH_2Br	H	NT	2.030	–	1.43
CH_2Cl	H	NT	2.448	–	2.85
C_6H_5	H	NT	0.190	–	1.5
CH_2OCH_3	H	NT	4.818	–	9.45
$c\text{-}C_6H_{11}$	H	NT	0.152	–	0.25
$CH_2CH=CH_2$	H	NT	0.487	–	0.95
NO_2	H	>0.75	>1.0	–	9.5
NH_2	H	>0.75	NT	–	>500
CH_3	CH_3	NT	12.3	–	55[d]
C_2H_5	CH_3	NT	1.183	–	3.15[d]
C_3H_7	CH_3	NT	0.102	–	0.75[d]
H	$t\text{-}C_4H_9$	NT	NT	–	38.5[d]
Picrotoxinin		NT	0.106	–	NT

a All values were corrected, using the mortality with 4-propyl BP (0.5 μg) alone as a standard.
b Synergistic ratio.
c Casida et al. (1976) and Eto et al. (1976); the values were corrected, using the LD_{50} value (0.18 mg/kg) of 4-isopropyl BP as a standard.
d Ozoe et al. (1983).
e Not tested.

Table 6. Regression analysis of structure-mammalian toxicity relationship for 4-substituted BPs

$R-C(CH_2O)_3P=O$ R	π [a]	E_s^c [c]	σ^* [e]	$Log(1/LD_{50}$ (mol/kg)) Obsd	$Log(1/LD_{50}$ (mol/kg)) Calcd
CH_3	0.66	0.00	0.00	3.71 [i]	3.79
C_2H_5	1.20	-0.38	-0.10	5.25 [i]	5.18
C_3H_7	1.74	-0.67	-0.12	5.70 [i]	5.73
$i-C_3H_7$	1.61	-1.08	-0.19	6.03 [i]	5.91
C_4H_9	2.28	-0.70	-0.13	5.14 [i]	5.22
$i-C_4H_9$	2.15 [b]	-1.24	-0.13	5.93	5.81
$s-C_4H_9$	2.15 [b]	-1.74	-0.21	5.99	6.05
$t-C_4H_9$	1.99	-2.46	-0.30	6.59	6.60
C_5H_{11}	2.82	-0.71	-0.13 [f]	3.77 [i]	3.80
$i-C_5H_{11}$	2.69 [b]	-0.66	-0.13 [g]	4.29	4.19
$CH(CH_3)C_3H_7$	2.69 [b]	-1.55 [d]	-0.21 [h]	4.70	4.71
$CH(C_2H_5)_2$	2.69 [b]	-2.59	-0.23	5.44	5.42
CH_2Br	0.51	-0.58	1.00	5.23	5.28
CH_2Cl	0.37	-0.55	1.05	4.84	4.79

a Calcd by use of the f value in Hansch and Leo (1979), unless otherwise noted.
b Estimated from the π value of each related substituent according to Hansch and Leo (1979).
c Taken from Fujita et al. (1973), unless otherwise noted.
d Calcd according to Eq. 19 in Fujita et al. (1973).
e Taken from Taft (1956), unless otherwise noted.
f Taken as that of n-butyl.
g Taken as that of isobutyl.
h Taken as that of s-butyl.
i Taken from Casida et al. (1976).

$$\log(1/LD_{50}\text{(mol/kg)}) = 5.230\ \pi - 1.541\ \pi^2 - 0.719\ E_s^c + 1.587\ \sigma^* + 1.006$$
$$(\pm 0.509)\quad (\pm 0.131)\quad (\pm 0.076)\quad (\pm 0.254)$$

$$n = 14 \quad r = 0.997 \quad s = 0.077 \quad F = 387.43$$

This equation was obtained by omission of some analogs from the data previously reported (Eto et al., 1976; Ozoe et al., 1981) and addition of new analogs to it. In the above equation, π is the hydrophobic parameter of substituents calculated based on the fragmental constants (Leo et al., 1975; Hansch and Leo, 1979), E_s^c Hancock's steric parameter (Hancock et al., 1961; Fujita et al., 1973), σ^* the inductive substituent constant for aliphatic substituents (Taft, 1956), n the number of compounds used for analysis, r the correlation coefficient, s the standard deviation, and F the value of the F-test. The above equation accounts for 99% ($R^2 = 0.99$) of the variance in the data. This finding indicates the

significance of electronic nature of the bridgehead substituent as
well as hydrophobic and steric ones. The toxicity is parabolically
related to the π constant, the optimum π value being 1.70. The
hydrophobic nature of the bridgehead substituent is important in
transport process of BPs to the site of action as well as in an
interaction of BPs with the site. The negative sign of the E_s^c term
indicates that the greater the bulkiness or branching of the
substituent, the higher the toxicity of BPs with the same hydrophobic
and electronic properties. The significance of σ^* indicates that an
electronic interaction of the substituent with a target site may be
involved in the mode of action of BPs.

When topically applied to the housefly, all BPs tested were
nontoxic at 15 µg/fly. Some BPs exerted toxicity, though weak, by
injection (Table 5). The most toxic BP was 4-propyl BP to houseflies, whereas 4-t-butyl BP was the most toxic to mice. The LD_{50}
values of these compounds for houseflies are equivalent to 21 µg/g
and 32 µg/g, which are 55 times and 603 times as large as those to
mice, respectively. The toxicity of BPs was potentiated by treatment
with the mixed-function oxidase inhibitor piperonyl butoxide (pb)
prior to the injection of BPs. In this case, the most toxic BP was
3-methyl-4-propyl BP comparable to the CABA antagonist picrotoxinin.
One of the strong contrasts to mammalian toxicity was the effect by
introduction of a methyl group into the C-3 position, which raised
insecticidal activity and lowered mammalian toxicity. Regression
analysis was applied to representative 4-substituted BPs listed in
Table 7. The best equation for insect (housefly) toxicity was

$$\log(1/LD_{50}(\text{mol/fly})) = 5.193\,\pi - 1.356\,\pi^2 - 0.316\,E_s^c + 1.711\,\sigma^* + 4.049.$$
$$(\pm 0.759)\ (\pm 0.207)\ \ (\pm 0.141)\ \ (\pm 0.370)$$

$$n = 12 \quad r = 0.992 \quad s = 0.108 \quad F = 112.86$$

The equation was similar to that of mammalian toxicity, indicating
the significance of each parameter of the substituent attached to the
bridgehead. A major difference between the above two equations is
that the coefficient of E_s^c in the former equation is approximately
twice as large as that in the latter. This finding and the effect by
the introduction of the 3-methyl group suggest the possibility that
the structure or nature of a critical target site is different
between the housefly and the mouse. However, caution is needed for
the interpretation of these findings, because the comparison is made
in two living systems in which BPs are transported and altered in a
different manner.

Speculation of a Critical Target Site

It was suggested in the preceding section that steric,
hydrophobic and electronic interactions between the bridgehead
substituent and its critical site of action play an important

Table 7.   Regression analysis of structure-housefly toxicity
relationship for 4-substituted BPs

$R-C(CH_2O)_3P=O$ R	π	E_s^c	σ^*	Log(1/LD$_{50}$(mol/fly)) Obsd	Calcd
CH_3	0.66	0.00	0.00	6.82	6.88
C_2H_5	1.20	-0.38	-0.10	8.29	8.28
C_3H_7	1.74	-0.67	-0.12	9.06	8.99
$i-C_3H_7$	1.61	-1.08	-0.19	9.01	8.92
C_4H_9	2.28	-0.70	-0.13	8.76	8.84
$i-C_4H_9$	2.15	-1.24	-0.13	9.09	9.12
$s-C_4H_9$	2.15	-1.74	-0.21	9.15	9.14
$t-C_4H_9$	1.99	-2.46	-0.30	9.22	9.28
$CH(CH_3)C_3H_7$	2.69	-1.55	-0.21	8.37	8.34
$CH(C_2H_5)_2$	2.69	-2.59	-0.23	8.64	8.63
CH_2Br	0.51	-0.58	1.00	8.08	8.24
CH_2Cl	0.37	-0.55	1.05	7.91	7.75

Ozoe et al. (1983)

role in the primary action of BPs. This situation may be
explained in terms of the receptor concept.

Cyclodextrin (CD), a cyclic oligosaccharide constructed from
six, seven or eight $(1 \to 4)$-linked glucose units, has a cylindrical
cavity in the molecule (Fig. 3) and includes a variety of
compounds which fit into the cavity to form inclusion complexes
(Griffiths and Bender, 1973). Hydrophobic and van der Waals
interactions are regarded as the major driving force for the
complexation (Matsui and Mochida, 1979). Examination of
interactions between 4-substituted BPs and CD may provide some
information on the structure or nature of the site of action or
of the receptor of toxic BPs, the CD being used as a simple model
of the receptor. The variation in the association constants (K_a)
for β-CD-BP complexes was quantitatively analyzed by use of
free-energy related physicochemical parameters and regression
equations (Ozoe et al., 1981). A regression equation was
obtained for 17 BPs listed in Table 8.

$$\log K_a (M^{-1}) = 0.068\,\pi - 0.189\,E_s^c + 2.415$$
$$(\pm 0.068)\quad(\pm 0.135)$$

$$n = 17 \qquad r = 0.826 \qquad s = 0.135 \qquad F = 15.04$$

This equation was considerably improved by omission of BPs with

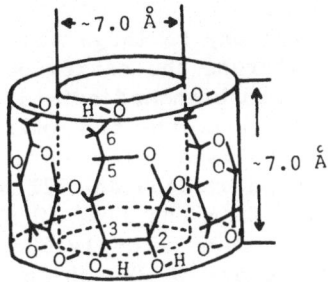

Fig. 3.   Structure of β-cyclodextrin.

heteroatom-substituted groups and 4-allyl BP (r = 0.897, s = 0.119).
The coefficient of the E_s^c term was negative whereas it was positive
in the case of α-CD-BP system. The negative sign indicates that the
inclusion complexes become stable with increasing bulkiness of the
substituent with a fixed π constant. The β-CD cavity (∿7.0 Å)

Table 8.   Substituent constants and association constants for
regression analysis

R-C(CH$_2$O)$_3$P=O R	π	E_s^c	Log K_a (M^{-1})	
			Obsd	Calcd
CH$_3$	0.66	0.00	2.31	2.46
C$_2$H$_5$	1.20	-0.38	2.44	2.57
C$_3$H$_7$	1.74	-0.67	2.53	2.66
i-C$_3$H$_7$	1.61	-1.08	2.77	2.73
C$_4$H$_9$	2.28	-0.70	2.65	2.70
i-C$_4$H$_9$	2.15	-1.24	2.76	2.80
s-C$_4$H$_9$	2.15	-1.74	2.79	2.89
t-C$_4$H$_9$	1.99	-2.46	2.98	3.02
C$_5$H$_{11}$	2.82	-0.71	2.61	2.74
i-C$_5$H$_{11}$	2.69	-0.66	2.97	2.72
ÇH(CH$_3$)C$_3$H$_7$	2.69	-1.55	2.99	2.89
OCH$_3$	-1.77	-0.05	2.38	2.30
CH$_2$Br	0.51	-0.58	2.61	2.56
CH$_2$Cl	0.37	-0.55	2.63	2.54
CH$_2$OCH$_3$	-0.46	-0.50	2.58	2.48
c-C$_6$H$_{11}$	2.99	-1.40	3.10	2.88
CH$_2$CH=CH$_2$	1.19	-1.48	2.63	2.78

Ozoe et al. (1981)

(Griffiths and Bender, 1973) may be large enough to enclose a bulky substituent sufficiently whereas the cavity of α-CD (4.5 Å) may be too small to include BPs with branched substituents deeply. Each BP has a symmetrical cage structure, the two ends of which are quite different in nature. One end is a polar phosphoryl group and the opposite one is a hydrophobic bridgehead. Thus BPs seem to enter deeply into the β-CD cavity from the side of the bridgehead, so that they can associate with β-CD more strongly than with α-CD. The most toxic analog 4-t-butyl BP is one of BPs which strongly associate with β-CD. Figure 4 shows a proposed mode of interaction of 4-t-butyl BP with β-CD. Taking into account the fact that the π^2 term in the equations of structure-toxicity relationship analysis described above is regarded as a term concerned with the passive penetration of bioactive compounds to the site of action, there is a similarity between the equations of structure-toxicity analysis and of the receptor model with regard to hydrophobic and steric parameters. If the inside diameter of a putative receptor is smaller than that of β-CD, i.e., if it resembles that·of α-CD, the sign of the E_s^c term in the equations of structure-toxicity analysis should be positive. On the contrary, if the inside of the receptor is much larger than the β-CD cavity, the E_s^c term may not contribute to the regression equation for lack of van der Waals interaction. Thus, these findings enable us to speculate that a pore-type receptor, the inside diameter of which is similar to that of β-CD (∿7.0 Å), is one of possible receptors for toxic BPs. Toxic BPs may fit into the pore-type receptor in such a way to block its normal physiological function.

Electrophysiological Studies

The toxic sign of BPs in both mice and houseflies, roughly speaking, resembles the general pattern of nerve poisoning: convulsive seizures, paralysis and hyperexcitability. However, the anticholinesterase activity of BPs was weak. Thus, it seems worth while examining the action of BPs against the nervous system by an electrophysiological method. In such a research on the nervous system, studies with simple synapses often provide fundamental and important information. The crustacean and annelida neuro-muscular junctions give good models for the synapses, in which γ-aminobutyric acid (GABA) acts as an inhibitory neurotransmitter (Takeuchi and Takeuchi, 1965, 1966; Toida et al., 1975). In the longitudinal

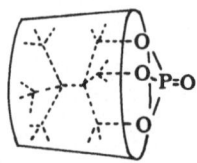

Fig. 4. A proposed mode of interaction of 4-t-butyl BP with β-CD.

muscle of the earthworm the inhibitory nerve innervation is more
dominant than the excitatory one, and the miniature inhibitory
junction potentials (m.i.j.p.s) can be recorded from nearly all the
cells in which microelectrodes are inserted (Hidaka et al., 1969; Ito
et al., 1969; Toida et al., 1975). The potential changes are due to
increased chloride conductance caused by the action of single GABA
quantum spontaneously released from the nerve terminals. This
section describes the effect of BPs on the spontaneously generated
inhibitory junction potentials recorded from the longitudinal muscle
of the earthworm (Korenaga et al., 1977).

Applications of 4-isopropyl BP (5 x 10^{-5} g/ml) induced
reduction in the amplitude and frequency of m.i.j.p.s. Figure 5
shows the amplitude histograms of m.i.j.p.s recorded from single cell
before and after the application of 4-isopropyl BP (10^{-4} g/ml). The
amplitude histogram of the m.i.j.p.s in the normal Ringer solution
was skew, indicating the generation of m.i.j.p.s. from the diffusely
innervated inhibitory nerve terminals. The mean amplitude of
m.i.j.p.s was 0.60 ± 0.38 mV (S.D.) and the total number was 211 in
50 seconds. After 20 minutes of the BP application, the amplitude
histogram shifted toward the small amplitude along the horizontal
direction, giving the mean amplitude value of 0.39 ± 0.20 mV and the
total m.i.j.p appearance was 89 in 50 seconds. The amplitude and
frequency of m.i.j.p.s progressively declined and the m.i.j.p.s.
practically disappeared after 40 minutes. The generation of
m.i.j.p.s was not restored even by one-hour rinsing with the normal
physiological solution.

In order to compare the potency of BPs as GABA antagonists with
that of picrotoxin, the frequency of m.i.j.p.s was measured on 10-30

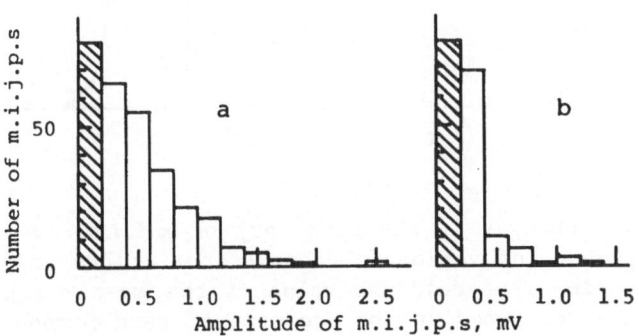

Fig. 5. Histograms of the amplitudes of the spontaneous miniature
         inhibitory junction potentials recorded from a single cell
         (a) before and (b) 20 minutes after the treatment with 4-
         isopropyl BP (10^{-4} g/ml). The histograms also show the
         total frequency of m.i.j.p.s observed in 50 seconds.

different fibers before and after treatment with these compounds.
The activities were compared with each other in terms of the ratio of
the mean frequency of m.i.j.p.s obtained in the presence of those
compounds against the values obtained in the physiological solution
(Fig. 6). Picrotoxin had only a small effect on the m.i.j.p.
frequency at 10^{-5} g/ml; i.e., in a series of experiments the mean
values of the m.i.j.p. frequency before and after the application of
picrotoxin was 22.3 $\pm$ 13.5 c/sec (n = 15) and 16.3 $\pm$ 10.7 c/sec (n =
20), respectively. However, at a concentration of 2 x 10^{-5} g/ml,
picrotoxin markedly reduced the frequency of m.i.j.p.s. The mean
reduction in three different preparations was 58.5 $\pm$ 4.8% (n = 3).

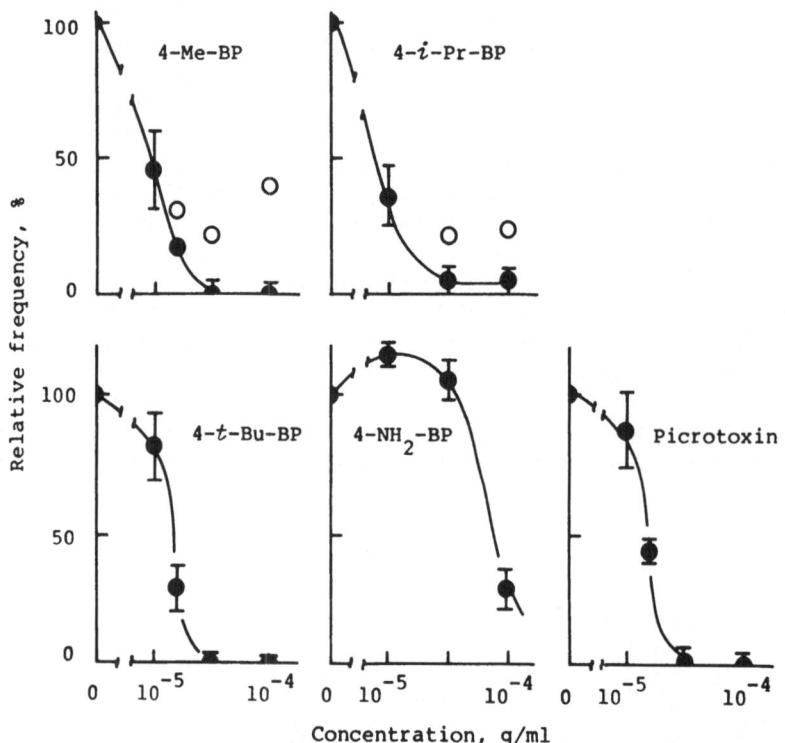

Fig. 6.   Relative changes in the frequency of m.i.j.p.s measured in
          the presence of various concentrations of BPs and
          picrotoxin. The relative value of the mean m.i.j.p
          frequency obtained in the presence of each compound against
          that obtained in the normal Ringer was plotted. Ten to
          thirty different fibers were used for the individual
          experiment. Each circle with a bar represents the mean $\pm$
          S.D. of 3-6 preparations: closed circle, the mean value of
          m.i.j.p. frequency during the application of chemicals; open
          circle, that after rinsing with the normal Ringer solution.

All alkyl BPs tested were effective for reduction in the m.i.j.p frequency at 10^{-5} g/ml. Especially 4-methyl and 4-isopropyl BPs showed strong actions on the size and frequency of m.i.j.p.s and reduced the mean frequency to less than 50% of the control value in one hour. At concentrations higher than 2×10^{-5} g/ml, all alkyl BPs decreased the mean frequency of m.i.j.p.s to less than 25% of the control value. Continuous washing-out procedure with the physiological solution for three hours did not restore the m.i.j.p. frequency to the control level. In contrast to the effect of alkyl BPs on m.i.j.p.s., 4-amino BP showed a rather weak action on the amplitude and frequency of m.i.j.p.s. At a concentration of 5×10^{-5} g/ml, which is enough for alkyl BPs and picrotoxin to induce the complete blockade of m.i.j.p. generation, 4-amino BP did not show any effect on the amplitude and frequency of m.i.j.p.s during the prolonged treatment up to three hours. At a high concentration (10^{-4} g/ml), 4-amino BP also reduced the amplitude and frequency of m.i.j.p.s. In addition, alkyl BPs slightly hyperpolarized the membrane in the concentration range 10^{-5} - 10^{-4} g/ml and reduced the frequency of spontaneously generated myogenic spike discharge at 10^{-6} and 10^{-5} g/ml.

On the other hand, the effect of BPs on the peripheral neuromuscular junction and the muscle was examined by use of the frog and the rat. The bicyclic phosphates did not elicit any effect on the amplitude or time course of the twitch contractions induced by nerve stimulation or direct muscle stimulation at concentrations up to 10^{-5} g/ml but 4-t-butyl and 4-isopropyl BPs suppressed the contractile activities of the frog and rat muscles at 10^{-4} g/ml.

Ligand-Receptor Binding Assays

The data described in the preceding sections suggested that there is a specific BP-binding site, which is closely related to the manifestation of toxicity, in the GABAergic synapses. The application of radioactive ligand binding assays with compounds with high specific activity makes it possible to investigate the postsynaptic mechanism at the subcellular level. Thus at first we tried to prepare radiolabeled BP analogs. 4-t-Butyl BP seems to be the best in terms of potency, but the product obtained was unsatisfactory in specific radioactivity because the synthetic procedure is more troublesome than those of other analogs. 4-[2,3-^{3}H]Propyl-2,6,7-trioxa-1-phosphabicyclo[2.2.2] octane 1-oxide ([^{3}H]4-propyl BP, 9.6 Ci) was instead prepared by the hydrogenation of 4-allyl BP (45 mg) with ^{3}H$_2$/Pt.

The binding of [^{3}H]4-propyl BP to rat brain membrane fractions (P_2 and P_3) was measured by a filter counting assay. As shown in Fig. 7, the amount of [^{3}H]4-propyl BP which bound to membrane preparations gradually decreased with increasing concentration of unlabeled 4-propyl BP in the incubation medium and approached to a

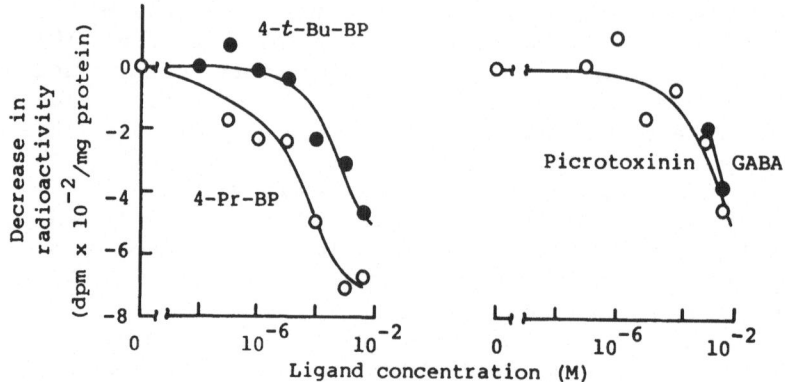

Fig. 7.   Displacement of [³H]4-propyl BP binding to rat brain
          synaptic membranes by BPs, GABA and picrotoxinin.

constant value at unlabeled BP concentrations above about 1 mM. This
finding indicates that [³H]4-propyl BP bound to a specific receptive
site is gradually displaced by unlabeled 4-propyl BP, which finally
saturates the site of BPs. Hence, the difference between the amounts
of bound [³H]4-propyl BP in the absence and presence of excess
unlabeled 4-propyl BP (1 mM) can be regarded as the amount of
specifically bound 4-propyl BP. The amount of specifically bound
[³H]4-propyl BP increased with an increase in the concentration of
[³H]4-propyl BP (Fig. 8), and the BP binding seemed to be a saturable
process, although the plots in Fig. 8 were somewhat scattered. The

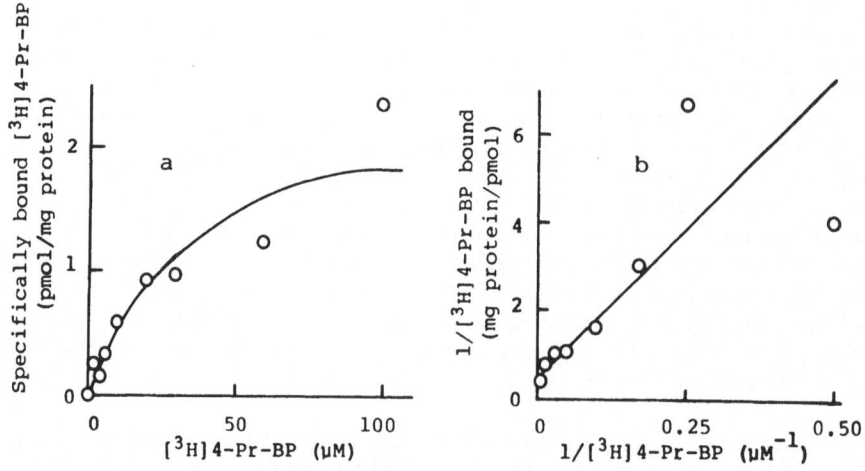

Fig. 8.   Specific binding of [³H]4-propyl BP to membranes from
          rat brain as a function of [³H]4-propyl BP concentration
          (a) and a double-reciprocal plot (b).

apparent equilibrium dissociation constant (K_d) and the quantity of
the BP bound at saturation (B_{max}) were estimated from the double-
reciprocal plot of the data to be about 30 uM and about 2 pmol/mg of
protein, respectively (Fig. 8). The potencies of various compounds
in displacing specifically bound [^{3}H]4-propyl BP were compared with
each other. Most BPs tested had the displacing activity. The only
exception was 4-amino BP, which was extremely low in toxicity (LD_{50} >
500 mg/kg) and had virtually no effect on specific [^{3}H]4-propyl BP
binding. However, no obvious correlation was observed between the
displacing potency for specific [^{3}H]4-propyl BP binding and the
toxicological potency of BPs (Table 9). 4-t-Butyl BP, the most toxic
compound among the convulsants of this class, was less active than 4-
propyl BP (Fig. 7 and Table 9). GABA and its noncompetitive
antagonist picrotoxinin also inhibited specific [^{3}H]4-propyl BP
binding at high concentrations (Fig. 7). On the contrary,
bicuculline methiodide, a stable derivative of the competitive GABA
antagonist bicuculline, characteristically enhanced the binding (Fig.
9). [^{3}H]4-propyl BP binding was profoundly influenced by sodium
chloride and the amount of specifically bound [^{3}H]4-propyl BP in the
absence of sodium chloride was approximately one-seventh of that in
the presence of 0.2 M sodium chloride (Fig. 10). The BP binding site
was extractable with the nonionic detergent Triton X-100.
Approximately 57% of specific binding was extracted into the
supernatant (100,000 x g, 30 min) by treatment with 0.05% (v/v) or
0.10 % (v/v) Triton X-100, although the binding activity was somewhat
lowered. The binding capacities of the supernatant and the pellet

Table 9. Displacement of [^{3}H]4-propyl BP bound to synaptic membranes
by unlabeled 4-substituted BPs and mammalian toxicity of
the BPs

Substituent	Decrease in the amount of [^{3}H]4-Pr-BP bound on addition of 5 mM unlabeled BP (dpm/mg pr.)	LD_{50} (mg/kg)
CH_3	410 ± 149	32
C_2H_5	490 ± 77	1.0
C_3H_7	666 ± 90	0.38
$i-C_3H_7$	350 ± 95	0.18
$t-C_4H_9$	467 ± 126	0.053
OCH_3	701 ± 137	8.0
CH_2OCH_3	621 ± 85	9.45
CH_2Br	432 ± 113	1.43
$CH(OH)CH_3$	469 ± 218	112.5
$COCH_3$	782 ± 274	51
NH_2	58 ± 78	>500

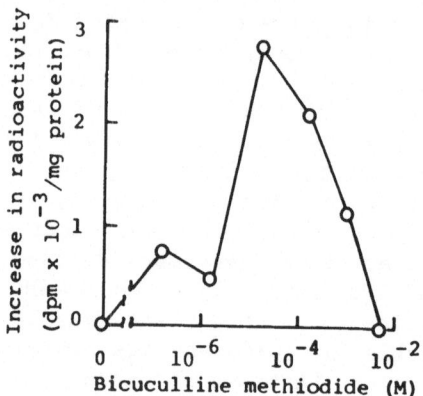

Fig. 9.   Effect of bicuculline methiodide on specific [^{3}H] 4-propyl BP binding to synaptic membranes.

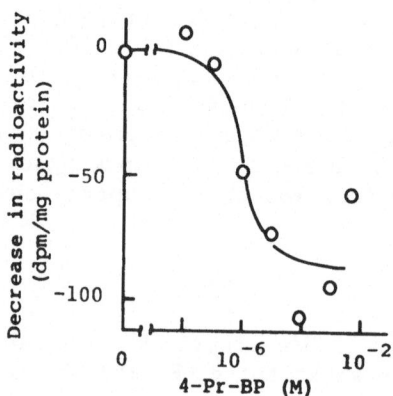

Fig. 10.   Displacement of [^{3}H]4-propyl BP binding by unlabeled 4-propyl BP in the absence of sodium chloride.

Table 10.   Effect of Triton X-100 on specific [^{3}H]4-propyl BP binding.

Triton X-100 (v/v %)	Specific binding of [^{3}H]4-Pr-BP (dpm/mg protein)	
	Supernatant	Pellet
0.05	486	363
0.10	298	226
0.10 [a]	440	146

a Binding activity was measured after storage at 4°C for 10 days.

obtained by treatment with Triton X-100 remained intact for at least 10 days under storage at 4°C (Table 10). The optimum temperature of specific [³H]4-propyl BP binding was approximately 20°C (Fig. 11). In addition, the following facts were observed (data not shown), although the data obtained by the filtration technique were considerably scattered. Specific [³H]4-propyl BP binding appeared to reach equilibrium within one minute and the binding was little affected by the pH of the incubation medium. The thermal stability of the binding site could not be determined since nonspecific binding was also enhanced by thermal treatment.

DISCUSSION

The bicyclic phosphates (analogs of 2,6,7-trioxa-1-phosphabicyclo[2.2.2] octane 1-oxide) are an interesting class of compounds in terms of structure, i.e., a symmetrical cage structure. More noticeable is the high toxicity to various animals (Casida et al., 1976; Kimmerle et al., 1976). The toxicity is strongly influenced by the physicochemical properties of the bridgehead substituent. About thirty BPs were synthesized in order to estimate the contribution of the substituent to the manifestation of toxicity quantitatively. Most BPs were prepared by cyclization between phosphoryl chloride and a triol. The phosphites and phosphoro-thionates can be prepared by similar reactions. A synthetic method of phosphorothionates from the corresponding phosphites and sulfur was useful for ³⁵S-labeling of the bicyclic phosphorus ester

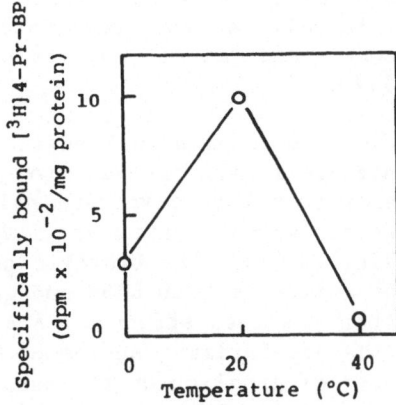

Fig. 11. Effect of temperature on specific [³H]4-propyl BP binding to synaptic membranes. The difference in radioactivity of [³H]4-propyl BP bound in the absence and presence of 1.0 mM unlabeled 4-propyl BP was regarded as specific [³H]4-propyl BP binding.

(Milbrath et al., 1979; Squires et al., 1983). The Hansch quantitative structure—activity relationship analysis suggested that the
toxic action of BPs occurred through an interaction between the
bridgehead substituent and a critical target site. It was
established that steric, hydrophobic and electronic properties of the
bridgehead substituents were significant for the interaction.
Sterical bulkiness of the substituent was more important in mammalian
toxicity than in toxicity to houseflies. The effect of the
substituents attached to the endomethylene carbon (C-3) was
impressive. Mammalian toxicity was lowered by the introduction of
the substituent whereas housefly toxicity was raised by it. This may
mean structural difference in the site of action between mice and
houseflies. Structure—activity relationship studies based on ligand-
receptor binding assays are necessary to draw a conclusion. As for
the other end of bicyclic phosphorus ester molecules, there is no
essential difference in toxicity among phosphites, phosphates and
phosphorothionates as far as the bridgehead substituent is the same.

     The bicyclic phosphates showed a relatively high reactivity
toward nucleophiles in an alkaline solution. Nevertheless, the
inhibition of acetylcholinesterases and α-chymotrypsin by BPs, which
proceeded by a mechanism similar to the above reaction was hardly
discernable except for weak inhibition by 4-nitro BP etc. Kinetic
analysis revealed that the low activity of BPs against the enzymes
was due to their low phosphorylating ability as well as their low
affinity. In contrast to other biologically active organophosphorus
esters, the phosphorylating ability of BPs was extremely weak. This
finding suggests that phosphorylation is not possibly involved in the
toxic action of BPs. It is conceivable that the endomethylene carbon
of BPs could be metabolically oxidized to produce the acyl phosphate
which may acylate a site of action so as to disturb its normal
physiological function. Actually an acyl phosphate (IV in Fig. 1)
was shown to acylate alcohols and amines, but was too unstable to be
toxic (Casida et al., 1976).

     It was shown by use of the longitudinal muscle of the earthworm
that BPs blocked the miniature inhibitory junction potentials caused
by spontaneous GABA release from the nerve terminals. The effect of
BPs on the mammalian central nervous system was investigated by
Bowery et al. (1976, 1977a, 1977b). The bicyclic phosphates
attenuated the depressant action of both GABA and glycine in the rat
medulla although the latter was less affected. Alkyl BPs dose-
dependently antagonized the depolarizing action of GABA observed in
the isolated superior cervical ganglion of the rat. 4-Ethyl and 4-
isopropyl BPs blocked presynaptic inhibition in the rat cuneate
nucleus (Davidson et al., 1977). 4-t-Butyl BP antagonized the GABA
inhibitory response of Limulus neurons but was much less active than
picotoxinin (Roberts et al., 1981). 4-Isopropyl and 4-t-butyl BPs
suppressed the twitch contractions induced by direct and indirect
peripheral nerve stimulation in the frog sartorious (Korenaga et al.,

1977). Thus enough electrophysiological evidence has accumulated
which shows that BPs act as GABA antagonists although they have not
been proved to be selective ones. Most of the actions are explained
by chloride ionophore blockade in the GABAergic system similarly to
that of the well-known antagonist picrotoxinin. The content of
cyclic GMP in the rat cerebellar cortex was significantly elevated by
injection of 4-isopropyl BP (Mattsson et al., 1977; Mattsson, 1980).
This seems to be associated with GABA inhibition although it may be
the secondary effect induced by a resulting increased release of the
excitatory transmitter. In this regard the inhibition of cyclic AMP
phosphodiesterase by 4-alkyl BPs was also reported (Coult and
Wilkinson, 1977). Bellet and Casida (1973) were the first to notice
structural similarity between the bicyclic phosphorus esters and
cyclic AMP. It is not clear to what extent interactions with the
cyclic AMP system contribute to the mode of action (Casida et al.,
1976).

The rat brain binding site of BPs was characterized by use of
$[^3H]$4-propyl BP as a ligand (Ozoe et al., 1982c). The affinity of
the ligand to the site was estimated to be lower than expected. The
low ratio of specific binding against total binding may be one of the
factors for it. The labeling of 4-t-butyl BP seemed to be better
than that of 4-propyl BP in terms of potency but $[^3H]$4-propyl BP was
selected because of easiness of labeling. The BP binding study was
extended by Squires et al. (1983) who overcame the difficulty of
preparation of the radiolabeled 4-t-butyl bicyclic phosphorus ester
with high specific activity by labeling with ^{35}S to produce $[^{35}S]$4-t-
butyl bicyclic phosphorothionate ($[^{35}S]$TBPS, I:R = t-C_4H_9, X = P=S in
Fig. 1). $[^{35}S]$TBPS bound with high affinity to specific sites in the
rat brain (K_d = 17 nM) and gave a desirable signal to noise ratio (67
: 33). A large increase in specific binding was attained by dialysis
of brain membrane fractions against an EDTA solution to remove
tightly bound GABA which works as an endogenous binding inhibitor.
This kind of treatment was not done in our preparation. This has
possibly resulted in a relatively large dissociation constant of 4-
propyl BP. Specific $[^{35}S]$TBPS binding was well characterized and
shown to be coupled to GABA-A and ion recognition sites.

It is widely accepted that the mammalian GABAergic transmission
system consists of at least three distinct components which are
allosterically coupled: GABA receptors, benzodiazepine recognition
sites and chloride ionophores. The natural toxin picrotoxinin
interferes with the ionic gating process by binding to a site related
to the ionophore. $[^{35}S]$TBPS binding was inhibited by picrotoxinin
and the reverse relation had been found to hold, i.e., the bicyclic
phosphorus esters inhibited the binding of a radiolabeled derivative
of picrotoxinin, $[^3H]$ α-dihydropicrotoxinin ($[^3H]$DHP]) to a specific
site (Ticku and Olsen, 1979). There is a possibility that the
bicyclic phosphorus esters and picrotoxinin share a site of action
(Hill et al., 1977). Toxicological findings as well as electro-

physiological ones support this possibility. Especially, the finding
that BPs showed cross-resistance to cyclodiene-resistant cockroaches
is important because it has been shown that the cyclodiene
insecticides act at the picrotoxinin binding site and that the
resistance phenomenon is caused by insensitivity of the site to those
compounds (Kadous et al., 1983).

However, there are still some questions whether BPs and
picrotoxinin act at the completely same site. The question is based
on several properties of $[^{35}S]TBPS$ binding which differ from those of
$[^3H]DHP$ binding although some of the differences are possibly due to
differences in properties of the ligands and other experimental
conditions. The most striking characteristic is the TBPS binding
inhibition by GABA and its mimetics. Specific $[^{35}S]TBPS$ binding was
potently inhibited by GABA and muscimol while $[^3H]DHP$ binding was not
(Ticku et al., 1978; Squires et al., 1983). The finding by Bowery et
al. (1984) further confused us that muscimol was a potent TBPS
binding inhibitor whereas GABA was not. In the case of $[^3H]$4-propyl
BP binding, the inhibitory effect of GABA was observed only at high
concentrations. Both $[^3H]$4-propyl BP and $[^{35}S]TBPS$ bindings were
greatly dependent on the concentration of sodium chloride and showed
maximum binding capacity at ca. 20°C. These findings are also
different from those of $[^3H]DHP$ binding (no dependence on sodium
chloride and optimum temperature, 0°C (Ticku et al., 1978)). High
sodium chloride concentration and maximum binding near 20°C imply
that a conformation change in the BP(TBPS)/picrotoxinin binding site
importantly alters the nature of binding of ligands to the site.
Multiplicity of $[^{35}S]TBPS$ binding (Squires et al., 1983) and
heterogeneity of $[^3H]DHP$ binding (Ticku, 1980; Davis and Ticku, 1981)
were also reported. Studies from the standpoint of membrane fluidity
appear to be necessary. It was speculated through model reaction
that the bridgehead substituents of toxic BPs fit into a pore-type
target site (Ozoe et al., 1981). It is controversial whether the
site is the chloride ionophore itself or the regulatory site of the
ionophore (Ticku and Maksay, 1983). There are two interesting
findings in this connection. The estimated diameter of the pore
(∿7 Å) is close to the diameter of the activated chloride ion channel
evaluated from the hydrated diameters of penetrating anion such as
bromate and chlorate (ca. 6 Å) (Araki et al., 1961). The hydrophobic
β-CD cavity used as a BP receptor model was shown to include anions
which penetrate the chloride channel (Mochida et al., 1973) as well
as toxic BPs.

Olsen et al. (1978) and Mattsson (1980) reported that BPs had no
effect on $[^3H]$GABA-receptor binding, whereas Enna et al. (1977)
obtained the opposite results that some BPs did inhibit the binding
under sodium-free conditions. More recent works established that the
inhibition of binding of $[^3H]$muscimol to the GABA receptor by BPs and
picrotoxinin occurred temperature-dependently in the presence of high
concentration of sodium chloride (Fujimoto and Okabayashi, 1981,

Karobath et al., 1981; Supavilia et al., 1982a, 1982b; Quast and Brenner, 1983). This finding is consistent with the dependence of specific bicyclic phosphorus ester binding on ions and indicates functional coupling of bicyclic phosphorus ester binding sites to GABA receptors. Uptake and release of GABA, and chloride uptake were not affected by 4-isopropyl BP (Olsen et al., 1978; Mattsson, 1980).

It is also interesting to note that the difference in [^{35}S]TBPS and [^{3}H]DHP binding inhibitory activities of the bicyclic orthocarboxylic acid ester (BOC) analogs (Table 11) (Ticku and Olsen, 1979; Squires et al., 1983). It was previously described that there were two optimum sizes of the substituents at the C-1 position of BOCs for high toxicity (Milbrath et al., 1979). One includes H and CH_3, and the other $n-C_4H_9$ and C_6H_5. The BOCs with the C-1 substituents of medium size had little toxicity. [^{35}S]TBPS binding was highly inhibited by the BOCs with the large C-1 substituents whereas the BOCs with the large C-1 substituents were less effective than those with small ones for inhibition of [^{3}H]DHP binding. It is conceivable that the site of action of the BOCs with the large C-1 substituents may be different from that of the BOCs with the small C-1 substituents or that the former site may overlap in part with the latter. Thus the BOC analogs with substituents of different size may recognize differences between TBPS and DHP binding sites. In similar respects, it may have to be considered that there is not evidence enough to decide whether the bicyclic phosphites, phosphates and phosphorothionates strictly act at the same site.

[^{35}S]TBPS is a good compound as a ligand in terms of specific to total binding ratio although it probably binds to the mixed-function oxidase etc. and the elimination of the sulfur atom easily occurs by

Table 11. Inhibition of specific [^{35}S]TBPS and [^{3}H]DHP bindings by bicyclic orthocarboxylic acid esters

$t-C_4H_9-C(CH_2O)_3C-R$ R	TBPS binding [a] IC_{50} (μM)	DHP binding [b] IC_{50} (μM)	LD_{50} [c] (mg/kg)
H	8.5	1 ± 1	6.5
CH_3	-	1.5 ± 1	6.0
C_2H_5	-	20 ± 8	80
$i-C_3H_7$	-	20 ± 6	>500
$n-C_3H_7$	-	10 ± 4	12
$n-C_4H_9$	0.059 ± 0.0038	25 ± 10	2.5
C_6H_5	0.035 ± 0.0012	40 ± 15	1.3

a Squires et al. (1983).
b Ticku and Olsen (1979).
c Casida et al. (1976) and Milbrath et al. (1978).

the action of the enzyme (Eto, 1974) (there is no problem as far as
specific binding is discussed), and is extensively used as an
alternative ligand for the picrotoxinin-related binding site. It was
reported that a variety of convulsant, depressant, anticonvulsant and
axiolytic drugs directly or allosterically interacted with the
binding site (Squires et al., 1983; Supavilai and Karobath, 1983;
Laurence et al., 1984; Ramanjaneyulu and Ticku, 1984a, 1984b;
Squires et al., 1984; Supavilai and Karobath, 1984; Ticku and
Ramanjaneyulu, 1984; Wong et al., 1984;).

Using [^{35}S]TBPS Lawrence and Casida (1983, 1984) reported that
the mode of action of α-cyano-3-phenoxybenzyl pyrethroids as well as
lindane, toxaphene and cyclodienes is associated with an interaction
between these compounds and the TBPS binding site in the rat brain.
These findings are of great interest from the point of view of
insecticide toxicology. The former compounds are one of classes of
the most promising insecticides and the latter chlorinated compounds
were major groups of insecticides whose mode of action had been
unknown for a long time. The latter compounds were shown to inhibit
[^{3}H]DHP binding to a specific site in the cockroach but the former
insecticides were not (Kadous et al., 1983; Matsumura and Ghiasuddin,
1983; Tanaka et al., 1984). [^{35}S]TBPS proved to be also a suitable
radioligand for autoradiographic study on the picrotoxinin site (Gee
et al., 1983). The TBPS (BP) binding site was recently solubilized
(Ozoe et al., 1982c; Seifert and Casida, 1983; King and Olsen, 1984;
Trifiletti et al., 1984) and the molecular weight was determined
(Nielsen and Braestrup, 1983). Labeled bicyclic phosphorus esters
are becoming important as a ligand for a specific site related to the
chloride ionophore in the GABA synapses. However, relation of the
ligand to the picrotoxinin binding site/the GABA receptor/the
benzodiazepine recognition site still remains to be solved at the
subcellular level.

REFERENCES

Araki, T., Ito, M., and Oscarsson, O., 1961, Anion permeability
     of the synaptic and nonsynaptic motoneurone membrane, J.
     Physiol., 159: 410-435.
Bellet, E.M., and Casida, J.E., 1973, Bicyclic phosphorus esters:
     high toxicity without cholinesterase inhibition, Science, 182:
     1135-1136.
Bowery, N.G., Brown, D.A., and Collins, J.F., 1975, Tetramethylenedi-
     sulphotetramine: an inhibitor of γ-aminobutyric acid induced
     depolarization of the isolated superior cervical ganglion of the
     rat, Br. J. Pharmacol., 53: 422-424.
Bowery, N.G., Collins, J.F., and Hill, R.G., 1976, Bicyclic
     phosphorus esters that are potent convulsants and GABA
     antagonists, Nature, 261: 601-603.

Bowery, N.G., Collins, J.F., Hill, R.G., and Pearson, S., 1977a, t-
    Butyl bicyclo phosphate: a convulsant and GABA antagonist more
    potent than bicuculline, Br. J. Pharmacol., 60: 275-276P.
Bowery, N.G., Collins, J.F., Hill, R.G., and Pearson, S., 1977b, GABA
    antagonism as a possible basis for the convulsant action of a
    series of bicyclic phosphorus esters, Br. J. Pharmacol., 60:
    435-346P.
Bowery, N.G., Hill, D.R., Hudson, A.L., Price, G.W., Turnbull, M.J.,
    and Wilkin, G.P., 1984a, Heterogeneity of mammalian GABA
    receptors, in: "Actions and Interactions of GABA and
    Benzodiazepines," N.G. Bowery, ed., Raven Press, New York.
Bowery, N.G., Price, G.W., Hudson, A.L., Hill, D.R., Wilkin, G.P.,
    and, Turnbull, M.J., 1984b, GABA receptor multiplicity.
    Visualization of different receptor types in the mammalian CNS,
    Neuropharmacology, 23: 219-231.
Casida, J.E., Eto, M., Moscioni, A.D., Engel, J.L., Milbrath, D.S.,
    and Verkade, J.G., 1976, Structure-toxicity relationship of
    2,6,7-trioxabicyclo[2.2.2]octanes and related compounds,
    Toxicol. Appl. Pharmacol., 36: 261-279.
Coult, D.B., and Wilkinson, R.G., 1977, The interaction of 4-alkyl
    derivatives of 2,6,7-trioxa-1-phosphabicyclo [2.2.2]octanes with
    cyclic adenosine 3',5'-monophosphate phosphodiesterase, and with
    cyclic adenosine 3',5'-monophosphate binding protein, Biochem.
    Pharmacol., 26:887-889.
Davidson, N., Macfarlane, E.I., and Michie, D.L., 1977, Comparison
    of bicyclic phosphorous esters with bicuculline and picrotoxin
    as antagonists of presynaptic inhibition in the rat cuneate
    nucleus, Experientia, 33:935-936.
David, W.C., and Ticku, M.K., 1981, Picrotoxinin and diazepam bind
    to two distinct proteins: further evidence that pentobarbital
    may act at the picrotoxinin site, J. Neurosci., 1:1036-1042.
Enna, S.J., Collins, J.F., and Snyder, S.H., 1977, Stereospecificity
    and structure-activity requirements of GABA receptor binding in
    rat brain, Brain Res., 124:185-190.
Eto, M., 1974, "Organophosphorus Pesticides: Organic and Biological
    Chemistry," CRC Press, Cleveland, Ohio.
Eto, M., Ozoe, Y., Fujita, T., and Casida, J.E., 1976, Significance
    of branched bridge-head substituent in toxicity of bicyclic
    phosphate esters, Agric. Biol. Chem., 40:2113-2115.
Fujita, T., Takayama, C., and Nakajima, M., 1973, The nature and
    composition of Taft-Hancock steric constants, J. Org. Chem.,
    38:1623-1630.
Fujimoto, M., and Okabayashi, T., 1981, Effect of picrotoxin on
    benzodiazepine receptors and GABA receptors with reference to
    the effect of Cl$^-$ ion, Life Sci., 28:895-901.
Gage, J.C., 1970, The subacute inhalation toxicity of 109 industrial
    chemicals, Br. J. Ind. Med., 27:1-18.
Gee, K.W., Wamsley, J.K., and Yamamura, H.I., 1983, Light microscopic
    autoradiographic identification of picrotoxinin/barbiturate
    binding sites in rat brain with [^{35}S]t-butyl-bicyclophosphoro-

thionate, _Eur. J. Pharmacol._, 89:323-234.

Griffiths, D.W., and Bender, M.L., 1973, Cycloamyloses as catalysts, _Adv. Catal._, 23:209-261.

Hancock, C.K., Meyers, E.A., and Yager, B.J., 1961, Quantitative separation of hyperconjugation effects from steric substituent constants, _J. Am. Chem. Soc._, 83:4211-4213.

Hansch, C., and Fujita, T., 1964, $\rho-\sigma-\pi$ Analysis. A method for the correlation of biological activity and chemical structure, _J. Am. Chem. Soc._, 86:1616-1626.

Hansch, C., and Leo, A., 1979, "Substituent Constants for Correlation Analysis in Chemistry and Biology," Wiley Intersciences, New York.

Hart, G.J., and O'Brien, R.D., 1973, Recording spectrophotometric method for determination of dissociation and phosphorylation constants for the inhibition of acetylcholinesterase by organophosphates in the presence of substrate, _Biochemistry_, 12:2940-2945.

Hidaka, T., Ito, Y., Kuriyama, H., and Tashiro, N., 1969, Neuromuscular transmission in the longitudinal layer of somatic muscle in the earthworm, _J. Exp. Biol._, 50:417-430.

Hill, R.G., Mitchell, J.F., and Pearson, S., 1977, Interactions of GABA antagonists on the isolated frog spinal cord, _Br. J. Pharmacol._, 61:484-485p.

Hollingworth, R.M., Fukuto, T.R., and Metcalf, R.L., 1967, Selectivity of Sumithion compared with methyl parathion. Influence of structure on anticholinesterase activity, _J. Agric. Food Chem._, 15:235-241.

Ito, Y., Kuriyama, H., and Tashiro, N., 1969, Effects of γ-aminobutyric acid and picrotoxin on the permeability of the longitudinal muscle of the earthworm to various anions, _J. Exp. Biol._, 51:363-375.

Kadous, A.A., Ghiasuddin, S.M., Matsumura, F., Scott, J.G., and Tanaka, K., 1983, Difference in the picrotoxinin receptor between the cyclodiene-resistant and susceptible strains of the German cockroach, _Pestic. Biochem. Physiol._, 19:157-166.

Karobath, M., Drexler, G., and Supavilai, P., 1981, Modulation by picrotoxin and IPTBO of ^{3}H-flunitrazepam binding to the GABA/benzodiazepine receptor complex of rat cerebellum, _Life Sci._, 28:307-313.

Kimmerle, G., Eben, A., Groning, P., and Thyssen, J., 1976, Acute toxicity of bicyclic phosphorus esters, _Arch. Toxicol._, 35:149-152.

King, R.G., and Olsen, R.W., 1984, Solubilization of convulsant/barbiturate binding activity on the -aminobutyric acid/benzodiazepine receptor complex, _Biochem. Biophys. Res. Commun._, 119:530-536.

Korenaga, S., Ito, Y., Ozoe, Y., and Eto, M., 1977, The effects of bicyclic phosphate esters on the invertebrate and vertebrate neuro-muscular junctions, _Comp. Biochem. Physiol._, 57C:95-100.

Large, W.A., 1975, Effect of tetramethylenedisulphotetramine on the membrane conductance increase produced by γ-amino-butyric acid at the crab neuromuscular junction, _Br. J. Pharmacol._, 53:598-599.

Lawrence, L.J., and Casida, J.E., 1983, Stereospecific action of pyrethroid insecticides on the γ-aminobutyric acid receptor-ionophore complex, Science, 211:1399–1401.

Lawrence, L.J., and Casida, J.E., 1984, Interactions of lindane, toxaphene and cyclodienes with brain-specific t-butylbicyclophosphorothionate receptor, Life Sci., 35:171–178.

Lawrence, L.J., Gee, K.W., and Yamamura, H.I., 1984, Benzodiazepine anticonvulsant action: γ-aminobutyric acid-dependent modulation of the chloride ioniphore, Biochem. Biophys. Res. Commun., 123:1130–1137.

Leo, A., Jow, P.Y.C., Silipo, C., and Hansch, C., 1975, Calculation of hydrophobic constants (Log P) from π and f constants, J. Med. Chem., 18:865–868.

Matsui, Y., and Mochida, K., 1979, Binding forces contributing to the association of cyclodextrin with alcohol in an aqueous solution, Bull. Chem. Soc. Jpn., 52:2808–2814.

Matsumura, F., and Ghiasuddin, S.M., 1983, Evidence for similarities between cyclodiene type insecticides and picrotoxinin in their action mechanisms, J. Environ. Sci. Health, B18:1–14.

Mattsson, H., 1980, Bicyclic phosphates increase the cyclic GMP level in rat cerebellum, presumably due to reduced GABA inhibition, Brain Res., 181:175–184.

Mattsson, H., Brandt, K., and Heilbronn, E., 1977, Bicyclic phosphorus esters increase the cyclic GMP level in rat cerebellum, Nature, 268:52–53.

Milbrath, D.S., Eto, M., and Casida, J.E., 1978, Distribution and metabolic fate in mammals of the potent convulsant and GABA antagonist t-butyl-bicyclophosphate and its methyl analog, Toxicol. Appl. Pharmacol., 46:411–420.

Milbrath, D.S., Engel, J.L., Verkade, J.G., and Casida, J.E., 1979, Structure-toxicity relationships of 1-substituted-4-alkyl-2,6,7-trioxabicyclo[2.2.2]octanes, Toxicol. Appl. Pharmacol., 47:287–293.

Mochida, K., Kagita, A., Matsui, Y., and Date, Y., 1973, Effects of inorganic salts on the dissociation of a complex of β-cyclodextrin with an azo dye in an aqueous solution, Bull. Chem. Soc. Jpn., 46:3703–3707.

Mochida, K., Matsui, Y., Ota, Y., Arakawa, K., and Date, Y., 1976, Substrate specificity in the cyclodextrin-catalyzed cleavage of organic phosphates and monothiophosphates in alkaline solutions, Bull. Chem. Soc. Jpn., 49:3119–3123.

Nielsen, M., and Braestrup, C., 1983, The molecular target size of brain TBPS binding sites, Eur. J. Pharmacol., 96:321–322.

Olsen, R.W., Ticku, M.K., Van Ness, P.C., and Greenlee, D., 1978, Effects of drugs on γ-aminobutyric acid receptors, uptake, release and synthesis in vitro. Brain Res., 139:277–294.

Ozoe, Y., Mochida, K., and Eto, M., 1981, Bicyclic phosphate binding to cyclodextrin: a receptor model, Agric. Biol. Chem., 45:2623–2625.

Ozoe, Y., and Eto, M., 1982a, Synthesis and some spectral character-
    istics of bicyclic phosphate GABA antagonists, Agric. Biol.
    Chem., 46:411-418.
Ozoe, Y., Mochida, K., and Eto, M., 1982b, Stability of bicyclic
    phosphates to alkali, Agric. Biol. Chem., 46:555-556.
Ozoe, Y., Mochida, K., and Eto, M., 1982c, Binding of toxic bicyclic
    phosphates to rat brain synaptic membrane fractions, Agric.
    Biol. Chem., 46:2521-2526.
Ozoe, Y., Mochida, K., and Eto, M., 1982d, Reaction of toxic bicyclic
    phosphates with acetylcholinesterases and α-chymotrypsin,
    Agric. Biol. Chem., 46:2527-2531.
Ozoe, Y., Mochida, K., Nakamura, T., Shimizu, A., and Eto, M., 1983,
    Toxicity of bicyclic phosphate GABA antagonists to the housefly,
    Musca domestica L. J. Pesticide Sci., 8:601-605.
Petajan, J.H., Voorhees, K.J., Packham, S.C., Baldwin, R.C., Einhorn,
    I.N., Grunnet, M.L., Dinger, B.G., and Birky, M.M., 1975,
    Extreme toxicity from combustion products of a fire-retarded
    polyurethane foam, Science, 187:742-744.
Quast, U., and Brenner, O., 1983, Modulation of [^{3}H]muscimol binding
    in rat cerebellar and cerebral cortical membranes by picrotoxin,
    pentobarbitone, and etomidate, J. Neurochem., 41:418-425
Ramanjaneyulu, R., and Ticku, M.K., 1984a, Binding characteristics
    and interactions of depressant drugs with [^{35}S]t-
    butylbicyclophosphorothionate, a ligand that binds to the
    picrotoxinin site, J. Neurochem., 42:211-229.
Ramanjaneyulu, R., and Ticku, M.K., 1984b, Interactions of penta-
    methylene-tetrazole and tetrazole analogues with the picro-
    toxinin site of the benzodiazepine-GABA receptor-ionophore
    complex, Eur. J. Pharmacol., 98:337-345.
Roberts, C.J., James, V.A., Collins, J.F., and Walker, R.J., 1981,
    The action of seven convulsants as antagonists of the GABA
    response of Limulus neurons, Comp. Biochem. Physiol., 70C:91-96.
Seifert, J., and Casida, J.E., 1983, Effect of solubilization of the
    t-butylbicyclophosphorothionate (TBPS) binding site on
    insecticide interactions with the GABA receptor-ionophore, 186th
    ACS Natl. Meeting, Division of Pesticide Chem., Abstr., 25.
Squires, R.F., Casida, J.E., Richardson, M., and Saederup, E., 1983,
    [^{35}S]t-Butylbicyclophosphorothionate binds with high affinity to
    brain-specific sites coupled to γ-aminobutyric acid-A and ion
    recognition sites, Mol. Pharmacol., 23:326-336.
Squires, R.F., Saederup, E., Crawley, J.N., Skolnick, P., and Paul,
    S.M., 1984, Convulsant potencies of tetrazoles are highly
    correlated with actions on GABA/ benzodiazepine/picrotoxinin
    receptor complexes in brain, Life Sci., 35:1439-1444.
Supavilai, P., Mannonen, A., and Karobath, M., 1982a, Modulation of
    GABA binding sites by CNS depressants and CNS convulsants,
    Neurochem. Internat., 4:259-268.
Supavilai, P., Mannonen, A., Collins, J.F., and Karobath, M., 1982b,
    Anion-dependent modulation of [^{3}H]muscimol binding and GABA-
    stimulated [^{3}H]flunitrazepam binding by picrotoxin and related

CNS convulsants, Eur. J. Pharmacol., 81:687-691.

Supavilai, P., and Karobath, M., 1983, Differential modulation of [^{35}S]TBPS binding by the occupancy of benzodiazepine receptors with its ligands, Eur. J. Pharmacol., 91:145-146.

Supavilai, P., and Karobath, M., 1984, [^{35}S]-t-Butylbicyclophosphoro-thionate binding sites are constituents of the γ-aminobutyric acid benzodiazepine receptor complex, J. Neurosci., 4:1193-1200.

Taft, R.M., Jr., 1956, Separation of Polar, Steric, and Resonance Effects in Reactivity, in: "Steric Effects in Organic Chemistry," Newman, M.S. ed., John Wiley & Sons, Inc., New York.

Takeuchi, A., and Takeuchi, N., 1965, Localized action of gamma-aminobutyric acid on the crayfish muscle, J. Physiol., 177:225-238.

Takeuchi, A., and Takeuchi, N., 1966, A study of the inhibitory action of γ-aminobutyric acid on neuromuscular transmission in the crayfish, J. Physiol., 183:418-432.

Tanaka, K., Scott, J.G., and Matsumura, F., 1984, Picrotoxinin receptor in the central nervous system of the American cockroach: its role in the action of cyclodiene-type insecticides, Pestic. Biochem. Physiol., 22:117-127.

Ticku, M.K., Ban, M., and Olsen, R.W., 1978, Binding of [^{3}H] γ-dihydropicrotoxinin, a γ-aminobutyric acid synaptic antagonist, to rat brain membranes, Mol. Pharmacol., 14:391-402.

Ticku, M.K. and Olsen, R.W. 1979, Cage convulsants inhibit picrotoxinin binding, Neuropharmacology, 18:315-318.

Ticku, M.K., 1981, Interaction of stereoisomers of barbiturates with [^{3}H] α-dihydropicrotoxinin binding sites, Brain Res., 211:127-133.

Ticku, M.K., and Maksay, G., 1983, Convulsant/depressant site of action at the allosteric benzodiazepine-GABA receptor-ionophore complex, Life Sci., 33:2363-2375.

Ticku, M.K., and Ramanjaneyulu, R., 1984, Ro 5-4864 inhibits the binding of [^{35}S]t-butylbicyclophosphorothionate to rat brain membranes, Life Sci., 34:631-638.

Toida, N., Kuriyama, H., Tashiro, N., and Ito, Y., 1975, Obliquely striated muscle, Physiol. Rev., 55:700-756.

Trifiletti, R.R., Snowman, A.M., and Snyder, S.H., 1984, Solubiliza-tion and anionic regulation of cerebral sedative/convulsant receptors labeled with [^{35}S]tert-butylbicyclophosphorothionate (TBPS), Biochem. Biophys. Res. Commun., 120:692-699.

Verkade, J.G., 1972/73, Spectroscopic studies of metal-phosphorus bonding in coordination complexes, Coordinat. Chem. Rev., 9:1-106.

Voorhees, K.J., Einhorn, I.N., Hileman, F.D., and Wojcik, L.H., 1975, The identification of a highly toxic bicyclophosphate in the combustion products of fire-retarded urethane foam, J. Polym. Sci., Polym. Lett. Ed., 13:293-297.

Wong, D.T., Threlkeld, P.G., Bymaster, F.P., and Squires, R.F., 1984, Saturable binding of ^{35}S-t-butylbicyclophosphorothionate to the sites linked to the GABA receptor and the interaction with GABAergic agents, Life Sci., 34:853-860.

THE GABA/BENZODIAZEPINE RECEPTOR·CHLORIDE CHANNEL: BIOCHEMICAL

IDENTIFICATION IN INSECTS AND STEREOSPECIFIC BINDING OF INSECTICIDES

Amira T. Eldefrawi, Ibrahim M. Abalis and Mohyee E. Eldefrawi

Department of Pharmacology and Experimental Therapeutics, University of Maryland School of Medicine, Baltimore, Maryland 21201

INTRODUCTION

γ-Aminobutyric acid (GABA) is the major inhibitory neurotransmitter in the central nervous system of vertebrates and invertebrates and also in invertebrate skeletal muscles. Only recently has it been shown to be a target for the toxic action of different kinds of insecticides.

There are at least three types of GABA receptors that differ in function, location and drug sensitivity. Activation of most GABA receptors leads to an influx (hyperpolarization) or efflux (depolarization) of Cl^-, causing in both cases inhibition, the type depending upon the neural localization of the receptor (postsynaptic or presynaptic). The direction in which Cl^- fluxes depends upon the Cl^- gradient across the membrane which is due mostly to changes in intracellular Cl^- concentration (see Enna and Gallagher, 1983). The GABA-activated Cl^- channels have been studied in mammalian spinal cord (McBurney and Barker, 1978), crustacean muscle (Onodera and Takeuchi, 1975) and insect muscle with a mean lifetime found in the latter to be 4-5 ms and conductance of approximately 9-23 pS at -80 mV (T = 21°C) (Cull-Candy, 1982). Differences in activities of GABA agonists are due to differences in channel lifetime and not unit conductance (Barker and Mathers, 1981; Gallagher et al., 1983). Most of these Cl^- channel-coupled receptors are $GABA_A$ receptors, which are activated by muscimol, inhibited by bicuculline and picrotoxinin (PTX) and occur mostly postsynaptically less so presynaptically. On the other hand, $GABA_B$ receptors are activated by baclofen, are insensitive to bicuculline, their GABA binding is dependent on Ca^{2+}, they are coupled possibly to a K^+ channel and occur presynaptically on nonGABAergic neurons (Bowery et al., 1981, 1983, Ong and Kerr, 1983; Enna, 1983).

GABA and benzodiazepine binding sites are coupled in the mammalian brain, since GABA responses are facilitated (Study and Barker, 1981) or inhibited (Steiner and Felix, 1976; Curtis et al., 1976) by benzodiazepines, and in vitro [^{3}H]GABA and [^{3}]flunitrazepam ([^{3}H]Flu) binding affinities are enhanced by diazepam and GABA, respectively (see Olsen, 1983; Squires, 1982). However, not all GABA and benzodiazepine binding sites are coupled, since their distributions in brain are not identical (Leeb-Lundberg et al., 1981). Also, the GABA$_B$ receptor is not coupled to a benzodiazepine receptor (Majewska and Chuang, 1984). The GABA and benzodiazepine binding sites are part of the same molecular entity, which was isolated in pure form very recently. There are also GABA$_B$ receptors, which do not have benzodiazepine binding sites, and there are low affinity benzodiazepine binding sites which regulate voltage-sensitive Ca^{+2} channels (Taft and DeLorenzo, 1984).

The ionic channel of the GABA$_A$ receptor is part of the receptor molecule (Fig. 1). There are inhibitory allosteric sites on the molecule, which are separate from the site that binds GABA and antagonists such as bicuculline, yet they inhibit GABA-induced Cl$^-$ conductances. These sites are assumed to be located closer to the receptor's channel component (also called "ionophore" though this implies incorrectly a molecule that shuttles ions across the membrane), and their action may in the future be found to differ according to the voltage across the membrane; thus they may be called

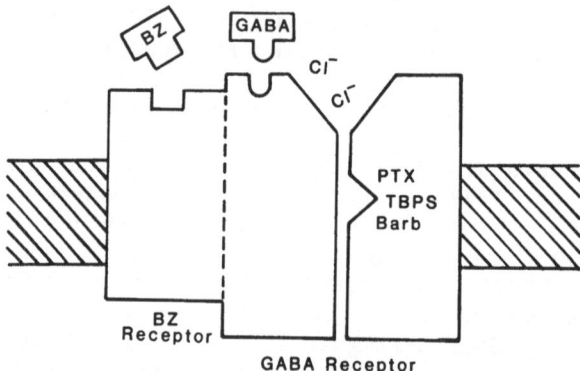

Fig. 1.   Schematic representation of a GABA$_A$ type receptor with its Cl$^-$ channel. It has two binding sites for GABA and one for benzociazepine (i.e., the BZ) receptor site. It also carries sites that are closely associated to the ionic channel: a site that binds picrotoxinin (PTX) and TBPS and possibly another for the sedative and hypnotic barbiturates (Barb.).

'channel sites.' The best known example is the site that binds the convulsant PTX, and inhibits both GABA-activated Cl^- flux and GABA enhancement of benzodiazepine binding without inhibiting [^{3}H]GABA (Olsen et al., 1978) or benzodiazepine (Leeb-Lundberg et al., 1981) binding. It also binds the hypnotic and sedative barbiturates (Ticku et al., 1978; Ticku and Maksay, 1983) as well as convulsants including bicyclic phosphates (Ticku and Olsen, 1979; Casida et al., 1976; Squires et al., 1983). This channel site in mammalian brain has been identified in vitro by its binding of [^{3}H]dihydropicro-toxinin ([^{3}H]DPTX) and more recently by [^{35}S]t-butylbicyclophorphoro-thionate ([^{35}S]TBPS), to which it has over 50-fold higher affinity (Squires et al., 1983). There may also be another channel site which binds pyrethroids since they inhibit noncompetitively [^{35}S]TBPS binding (Lawrence and Casida, 1983).

REVIEW OF RELATED WORK

Although most of the available electrophysiological information on GABA receptor function (e.g., channel lifetime and conductance) is on the accessible invertebrate inhibitory neuromuscular junction, biochemical identification of GABA receptors in insects has lagged far behind that of mammalian brain. In fact, until recently it was assumed that invertebrates do not have benzodiazepine receptors since specific binding could not be detected (Nielsen et al., 1978). This is due to reduced interest and to additional problems that one faces in biochemical studies of insect receptors. The techniques used in mammals cannot simply be transposed to insects. In order to obtain enough tissue from insects, the preparation is often contaminated with different organs and more proteolytic enzymes; thus it is necessary to add more antiproteases (Eldefrawi et al., 1982). An appropriate probe for a mammalian brain receptor site such as TBPS may not be so for the insect receptor. Finally, insect species may differ, such as our detection of reasonable binding of [^{3}H]Flu to house fly muscle, but less so to honey bee muscle, yet detection of reasonable binding of [^{3}H]muscimol to honey bee brain (unpublished information).

Earlier research suggested that GABA receptors may be molecular targets for insecticides. Several benzodiazepines were found to be toxic to aphids though not to the house fly, German cockroach or the confused flour beetle (Clifford & Jeffrey, 1977). PTX analogs were potent convulsants in desheathed house fly thoracic ganglion (Miller et al., 1979), but had little toxicity to house flies (Kuwano et al., 1980). In addition, cockroaches resistant to lindane and dieldrin (Kadous et al., 1983) were found resistant to PTX also, suggesting that they may have a common molecular target.

The first direct evidence for the action of insecticides on GABA receptors was provided by Dr. Matsumura, who showed that heptachlor epoxide and lindane inhibited [^{3}]DPTX binding to rat brain membranes and GABA-induced $^{36}Cl^-$ uptake by cockroach muscle (Ghiasuddin and Matsumura, 1982; Matsumura and Ghiasuddin, 1983).

Another line of evidence was provided by Dr. Casida and collaborators. First, several bicyclophosphate convulsants were synthesized and found to be very toxic to mammals (Casida et al., 1976; Milbrath et al., 1978). Then TBPS was radiolabeled with ^{35}S and found to have higher affinity than [^{3}H]DPTX for its binding site, a K_d of 17 nM compared to 1 μM (Squires et al., 1983). Studies of the effect of the two types of pyrethroids on [^{35}S]TBPS binding to rat brain synaptosomes showed that it was inhibited by 37 α-cyano-3-phenoxybenzyl pyrethroids with an absolute correlation with their mouse intracerebral toxicity (Lawrence and Casida, 1983). The non-cyano-containing pyrethroids had little or no effect. This led to the suggestion that the mechanism of toxicity of the cyanophenoxy-benzyl pyrethroids may involve an interaction with the GABA receptor complex. In a very recent study of 34 chlorinated hydrocarbon insecticides, concurrent with our own, Lawrence and Casida (1984) reported that lindane, toxaphene and cyclodienes bound stereo-specifically to the [^{35}S]TBPS binding site in rat brain, with potencies related to their mammalian toxicities, acting as non-competitive GABA antagonists. On the other hand, DDT, mirex and kepone at 10 μM did not inhibit [^{35}S]TBPS binding.

Avermectin B_{1a}, the macrocyclic anthelmintic drug which also has some insecticidal activity (Ostlind et al., 1979), was found to increase Cl^- conductance in lobster muscle, an effect that was blocked by picrotoxin (Fritz et al., 1979). This action was due to its binding to the GABA/benzodiazepine receptor, though at an allosteric site, for it potentiated [^{3}H]Flu and [^{3}H]GABA binding to rat brain mambranes and increased the maximal number of binding sites for both [^{3}H]Flu (Williams and Yarbrough, 1979; Supavilai and Karobath, 1981) and [^{3}H]GABA (Pong and Wang, 1982). Only its effect on [^{3}H]GABA binding was inhibited by picrotoxin (Pong et al., 1981). Though avermectin B_{1a} was like pentobarbital in stimulation of [^{3}H]Flu binding being additive to the stimulation produced by GABA, avermectin B_{1a} was different (also from picrotoxin) in protecting [^{3}H]Flu and [^{3}H]GABA binding against heat inactivation (Pong et al., 1982). Thus, it was suggested that avermectin B_{1a} binds to the GABA/benzodiazepine receptor at a site that may be partially shared by picrotoxin and pentobarbital (Pong et al., 1982). We have found [^{35}S]TBPS binding to rat brain membranes was inhibited avermectin B_{1a} as detailed below.

EXPERIMENTAL

In Vitro Identification of a GABA Receptor in Insects

House fly thoraces were isolated by differential sieving of dry-ice-frozen adults, homogenized in Van Harreveld's buffer (which contains high Ca^{2+} and Mg^{2+} concentrations) (1936), filtered over cheesecloth (to remove cuticular debris), centrifuged at 100,000 x g for 30 min and the supernatant was discarded to remove proteases and possible soluble endogenous inhibitors (Abalis et al., 1983). The pellet was rehomogenized in same buffer and the supernatant of a 1000 x g for 10 min centrifugation, recentrifuged at 100,000 x g for 30 min then the pellet resuspended in 50 mM Tris-citrate buffer (pH 7.1) (∿20 mg protein/ml).

The benzodiazepine receptor was identified by its specific binding of [^{3}H]Flu (98.3 Ci/mmol, New England Nuclear), which was measured by filtration after a 30-min incubation at $4^{o}C$ on Whatman GF/B filters (presoaked in 1% Prosil-28 (PCR Research Chemicals) to reduce nonspecific binding to the filter). Each filter was washed with 10 ml ice-cold 50 mM Tris-citrate buffer to remove all unbound ligand, then the filter with the membranes was placed in 4 ml toluene-based scintillation solution, and the radioactivity counted after at least 8 h. Nonspecific binding was defined as that occurring in presence of 50 μM diazepam.

Interactions of Insecticides with GABA Receptors of Rat Brain

Rat brains were homogenized in 20 vol of ice-cold 0.32 M sucrose, 1 mM EDTA, 1 mM Tris-HCl buffer, pH 7.1, and the supernatant of a 1000 x g centrifugation for 10 min was recentrifuged at 48,000 x g for 30 min. The pellet was stored overnight at $-90^{o}C$, resuspended in 50 mM Tris 1 mM EDTA (pH 7.1), centrifuged (after 20 min at $22^{o}C$) at 48,000 x g for 30 min and the step repeated 3 times. The final membrane preparation was in 5 mM Tris, 1 mM EDTA buffer pH 7.1) at 5 mg protein/ml and stored at $-90^{o}C$.

Binding of [^{3}H]Flu to the benzodiazepine receptor site was measured after 90 min incubation of the membranes at $0^{o}C$ in 50 mM Tris-citrate buffer, pH 7.1, by filtration on Whatman GF/B filter and counting filter radioactivity as described above. Nonspecific binding was that in presence of 3 μM diazepam. Binding of [^{3}H]muscimol (7.2 Ci/mmole, New England Nuclear) to the GABA receptor site was measured after 30 min incubation with the membranes at $0^{o}C$ in 50 mM Tris-HCL (pH 7.1) also by filtration on Whatman GF/B filter. Nonspecific binding was that in presence of 20 μM GABA. Binding to the channel site was measured by incubating the membranes for 90 min at $21^{o}C$ with 2 nM [^{35}S]TBPS in presence of 200 nM KBr, 1 mM EDTA, 5 mM Tris-HCl buffer, pH 7.5 (Squires et al., 1983), then filtering them on GF/B filters as described above. Nonspecific binding was that in presence of 20 μM PTX.

In Vitro Identification of GABA Receptors in Insects

Using either [3H]GABA or [3H]m scimol to identify GABA receptors in house fly thoraces was unsuccessful because of the high variability and low percentage of specific binding. On the other hand, significant amounts of specific [3H]Flu binding was detected, which was dependent upon protein (Fig. 2) and ligand (Fig. 3) concentrations. Two binding affinities were detected (Fig. 4): a K_d of 24.3 nM to 0.2 pmol/mg protein and a K_d of 1 μM to 8.2 pmol/mg protein (Abalis et al., 1983). These are lower affinities than that of 2.8 nM reported for human brain cortex (Speth et al., 1978) or of 1 nM for rat brain (Braestrup and Squires, 1978). The observed rate constant of association (k_{obs}) was calculated to be 6.2×10^{-3} s^{-1} for the slow component in Fig. 5, which represented the majority of points. The K_d value was calculated from the dissociation rate constant ($\underline{k}_{-1}$) obtained by measurement of binding after addition of 50 μM diazepam (5.4×10^{-3} s^{-1}) and association rate constant ($\underline{k}_{+1}$) calculated as 3.2×10^{5} $^{-1}$ s^{-1} from the equation $\underline{k}_{obs} - \underline{k}_{-1} = \underline{k}_{+1}$.

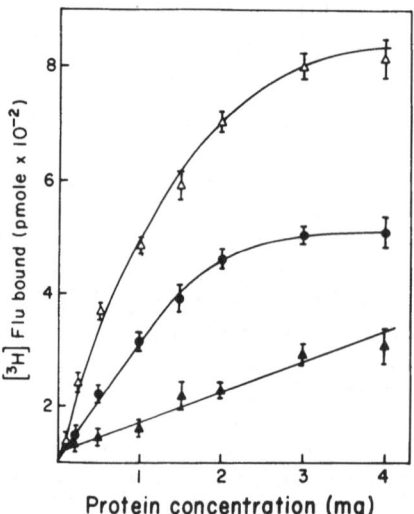

Fig. 2. [3H]Flu binding to membranes from house fly thoraces, in 50 mM Tris-citrate buffer, pH 7.1, as a function of protein concentration. Incubation volume was 1 ml, [3H]Flu concentration 2.5 nM, time was 30 min and temperature 0.4°C. Total binding ($\triangle$), non-specific binding ($\blacktriangle$) (measured in presence of 50 m diazepam), and specific binding ($\bullet$) (the difference between total and nonspecific binding). Each symbol and vertical bar is the mean of triplicate experiments $\pm$ SD (from Abalis et al., 1983).

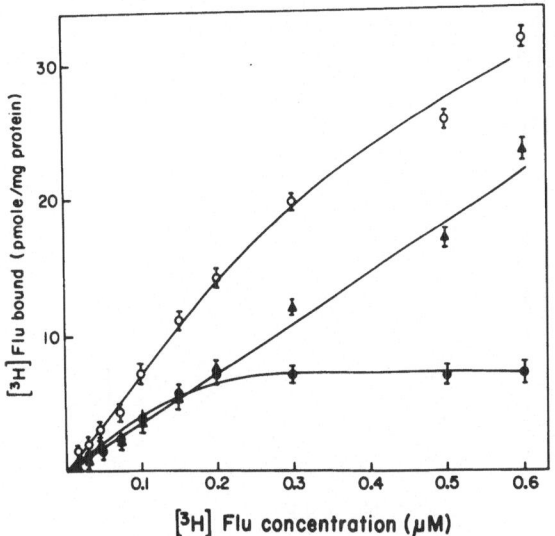

Fig. 3. Saturation isotherm of [3H]Flu binding to membranes from
house fly thoraces. [3H]Flu was diluted with unlabeled Flu
to obtain concentrations higher than 1 nM. The final concentrations ranged from 1 nM and 1 μM. Total binding (O),
nonspecific binding (▲) measured in presence of 50 M
diazepam, and specific [3H]Flu binding (●). Each symbol and
vertical bar is the mean + SD of three experiments (from
Abalis et al., 1983).

[3H]Flu. The calculated K_d of 16.8 nM is close to that of 24.3 nM
for the high affinity binding obtained (Fig. 4).

Flu and diazepam were more potent inhibitors of [3H]Flu binding
at 2.5 nM than the benzodiazepine antagonist ethyl- β -carboline-3-
carboxylate (β-CCE) and the agonists Ro 5-3027, Ro 5-2180, Ro 5-4864
(Table 1). The low potency of clonazepam and high potency of Ro 5-
4864 were similar to their potencies on [3H]Flu binding to mammalian
brain (Speth et al., 1978; Braestrup et al., 1981). It should be
noted that [3H]Flu binding at 2.5 nM represents binding to both the
high affinity (or receptor site) and low affinity binding sites,
about 60% to the former and 40% to the latter. It explains why the
apparent K_i value for Flu (Table 1) is about 10X higher than K_d of
24.3 nM for [3H]Flu binding. By analogy, the apparent K_i values for
the other benzodiazepines may also be about 10X higher than their
actual K_d. The data suggest that the benzodiazepine receptor in
house fly muscle is closer to those in peripheral tissues than brains
of mammals.

Like mammalian brain receptors, [3H]Flu binding to the house fly
membranes was stimulated in a dose-dependent manner by GABA and

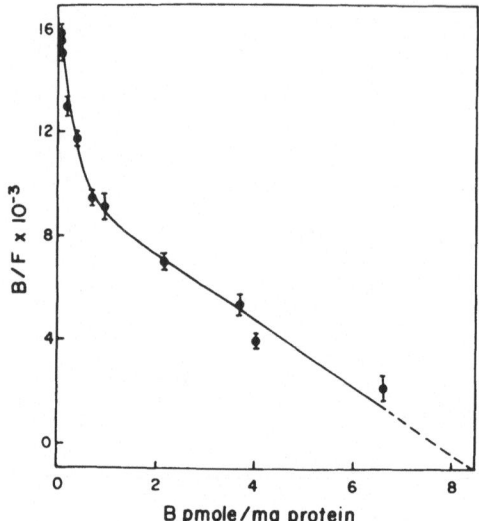

Fig. 4. Scatchard plot of the [3H]Flu specific binding to membranes
         from house fly thorax (Fig. 3). B, represents bound [3H]Flu
         in picomoles per milligram protein, and F is the concentra-
         tion of free ligand (i.e., [3H]Flu) in nM. Each symbol and
         vertical bar is the mean ± SD of three experiments (from
         Abalis et al., 1983).

agonists (muscimol and imidazole acetic acid) but not the GABA uptake
inhibitor DL-2,4-diaminobutyric acid. Binding of the benzodiazepine
site on the GABA receptors of mammalian brain was also not affected

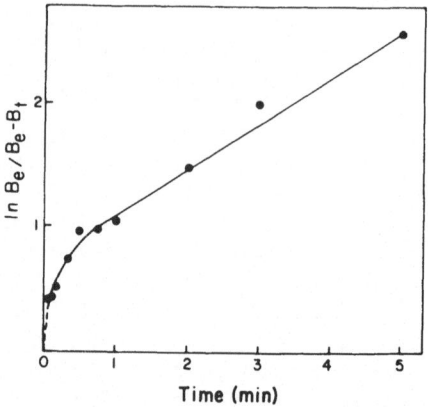

Fig. 5. Linear transformation of the time dependence of specific
         binding of [3H]Flu to membranes from house fly thorax.
         Incubation mixtures contained 1 mg membrane protein and 2.5
         nM [3H]Flu. Each symbol and vertical bar is the mean ± SD
         of three experiments (from Abalis et al., 1983).

Table 1. Comparison of benzodiazepine binding sites in different
species and tissues (from Abalis et al., 1983)

Compound	K_i (nM)		
	House fly thorax[a]	Mammalian brain	Rat kidney
Flunitrazepam	290 $\pm$ 35	2.72[b]	---
Diazepam	488 $\pm$ 60	27.4[b]	---
Clonazepam	146,000 $\pm$ 3,522	1.13[b]	1,790[d]
Ro 5-3027	9,800 $\pm$ 605	1.24[b]; 4.4[c]	---
Ro 5-2180	9,800 $\pm$ 460	8.8[c]	---
Ro 5-4864	680 $\pm$ 75	100,000[d]	2.9[d]
β-CCE	9,800 $\pm$ 325	1.13[e]	---

[a]Each value is the mean of three separate experiments, performed in
triplicate, $\pm$ standard deviation. It is an apparent K_i, since
[^{3}H]Flu binding at 2.5 nM represents binding to about 60% of
benzodiazepine binding sites and 40% to low affinity sites.
[b]Data on [^{3}H]Flu binding to human cerebral cortex membranes from
Speth et al. (1978).
[c]Data on [^{3}H]diazepam binding to rat brain membranes from
Braestrup and Squires (1978).
[d]Data calculated from IC_{50} values of [^{3}H]diazepam binding to rat
tissues from Braestrup and Squires (1977).
[e]Data calculated from IC_{50} values of [^{3}H]Flu binding to rat
cerebellum from Braestrup et al. (1981).

by all GABA agonists (Braestrup et al., 1980; Ferkany et al., 1981).
For example, GABA and muscimol potentiated binding, while piperidine-
4-sulfonic acid had no effect. Imidazole acetic acid, which was
potent in potentiating binding of [^{3}H]Flu to the house fly receptor,
had no effect on the mammalian brain receptor. The GABA antagonist
bicuculline (10 nM - 20 μM) also potentiated [^{3}H]Flu binding up to
50% and inhibited it at higher concentrations (Fig. 6) as was
reported for rat brain (Karobath et al., 1981). The reduced potency
of high concentrations of GABA, bicuculline or picrotoxinin to
stimulate [^{3}H]Flu binding (Figs. 6,7) may be related to the
phenomenon of receptor desensitization or fading of response
(Gallagher et al., 1983). Since at 17 μM bicuculline had no effect
on [^{3}H]Flu binding, we used this concentration to determine if
bicuculline inhibited the stimulation by GABA of [^{3}H]Flu binding,
which it did (Fig. 7). This allosteric interaction suggests that

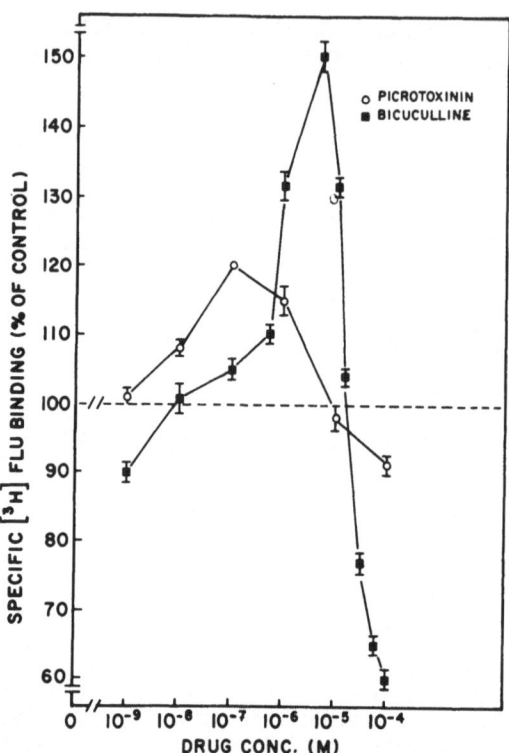

Fig. 6. The effect of the two classical GABA antagonists,
bicuculline and picrotoxinin on [³H]Flu binding to house
fly thorax-muscle membranes. Incubation was at 0-4°C
for 30 min in 5 mM Tris-HCl, pH 7.1. Bicuculline stock
solution of 100 µM at pH 3.2 2as kept at -20°C, and
picrotoxinin stock solution was in ethanol. The results
are mean values of triplicate determinations, each
performed in triplicate.

[³H]Flu binds to a benzodiazepine receptor in the house fly, which is
coupled to a GABA receptor. This coupling, and the bicuculline
effect, suggest that it is a $GABA_A$ type receptor. Since skeletal
muscles make up >90% of the tissues in house fly thorax, and we could
not detect significant high affinity binding of [³H]Flu to house fly
head preparation, it is likely that the observed binding to thoracic
membranes is to skeletal muscles.

Interactions of GABA Receptors with Insecticides

Since the GABA/benzodiazepine receptor complex carries different
ligand binding sites, we developed a rat brain preparation in which
which we could study binding of [³H]muscimol (4 nM) to the GABA
receptor site, [³H]Flu (1 nM) to the benzodiazepine binding site and

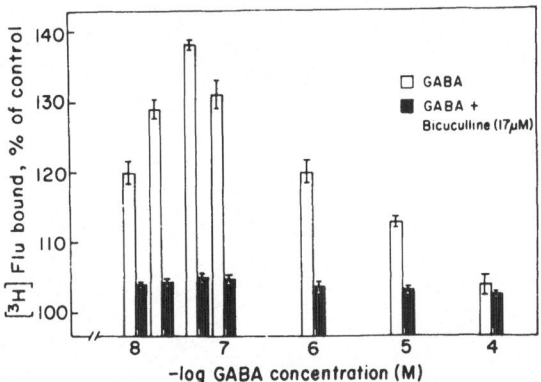

Fig. 7.   The stimulation by GABA of 2.5 nM [³H]Flu binding to membranes from house fly thorax in 5 mM Tris-Hcl buffer, pH
7.1, and the inhibition of GABA effect by 17 μM bicuculline.
Control 100% level represents the specific binding in
absence of GABA.   Open columns represent the specific [³H]
Flu binding in presence of the indicated concentration of
GABA.   Solid columns represent binding in presence of both
GABA and 17 μM bicuculline.   Vertical bars represent
standard deviation of four experiments (from Abalis et al.,
1983).

[³⁵S]TBPS (2 nM) to the channel sites.   [³H]Flu binding was potentiated by GABA, while [³H]muscimol binding was inhibited by GABA, but
potentiated by diazepam, and [³⁵S]TBPS binding was inhibited by GABA,
suggesting allosteric interactions between the three kinds of binding
sites as previously reported (Olsen, 1983; Squires et al., 1983).
Since the compounds used were solubilized in ethanol, all controls
contained the same amount of alcohol as the treated samples.   Final
alcohol concentration was 1% in assaying [³H]Flu and [³H]muscimol
binding, but was 2% in assaying [³⁵S]TBPS binding.   The K_d value of
45 nM that we obtained for specific [³⁵S]TBPS binding by dilution
with unlabeled TBPS (Abalis et al., in press) was similar to those of
17 nM (Squires et al., 1983) and 66 nM (Lawrence and Casida, 1983)
previously reported to rat brain membranes.

Binding of [³H]Flu (at 1 nM) and [³H]muscimol (at 4 nM) to rat
brain or house fly muscle membranes was unaffected by γ-hexachlor-
ocyclohexane (γ-BHC), aldrin, dieldrin, heptachlor or heptachlor
epoxide at 1 to 10,000 nM (Abalis et al., in press).   However, these
insecticides inhibited [³⁵S)TBPS binding to rat brain membranes in a
concentration-dependent manner (Fig. 8) with the following descending
order of potency: endrin = endosulfan I (K_i 30 nM) > endosulfan II
(60 nM) > heptachlor epoxide (70 nM) > dieldrin (100 nM) > γ-BHC
(150 nM) > heptachlor (400 nM) > aldrin (500 nM).   Hexachlorobenzene
and γ-BHC (400 nM) had no effect.   The inhibition by endrin of

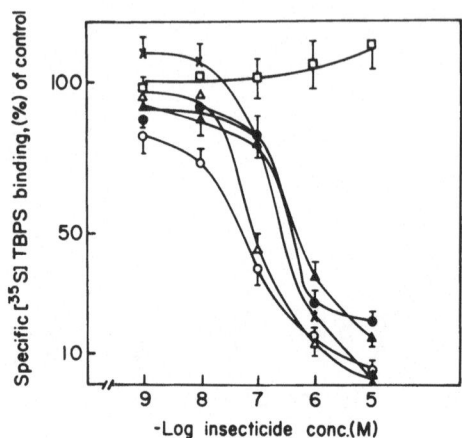

Fig. 8.   Inhibition of the specific binding of [^{35}S]TBPS (2 nM) to
          GABA receptors in rat brain membranes by picrotoxinin (x),
          two cyclodiene insecticides (heptachlor (●) and aldrin (▲)),
          their epoxides heptachlor epoxide (o) and dieldrin (△),
          respectively, and hexachlorobenzene (□). All chemicals were
          added to the incubation buffer in 10 μl aliquots prior to
          addition of tissue. Final volume of incubation was 1 ml of
          KBr-Tris-EDTA buffer, incubation time was 90 min at 21°C and
          20 mg of original brain tissue were used per assay.
          Nonspecific binding was determined in presence of 20 μM
          picrotoxinin (from Abalis et al., in press).

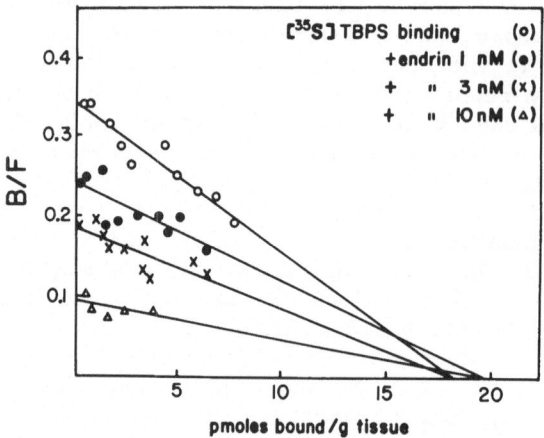

Fig. 9.   Scatchard plot of the binding of [^{35}S]TBPS to the ionic
          channel site of the GABA receptor of rat brain in absence
          (o) and presence of 1 nM (●), 3 nM (x) and 10 nM (△) endrin.
          Data were analyzed by regression analysis. B, amount bound
          in pmoles/g tissue; F, free [^{35}S]TBPS binding 51 nM and
          maximum number of binding sites is 19 $\pm$ 2 pmol/g tissue
          (from Abalis et al., in press).

[^{35}S)TBPS binding to rat brain was studied in more detail and found to be competitive as shown by the increase in K_d without affecting maximal binding (Fig. 9). Since we could not detect specific [^{35}S]TBPS binding to the house fly preparation, we have not yet studied the effects of these insecticides on the ionic channel component of the house fly GABA receptor. The effect of these insecticides on rat brain was stereospecific; the more toxic compound was more potent in inhibiting [^{35}S]TBPS binding (Fig. 10), and the epoxide derivative were more potent as reported by Lawrence and Casida (1984).

Another insecticide that was suggested to be interacting allosterically with the GABA/benzodiazepine receptor complex was avermectin B_{1a}; therefore we tested it, along with two other mycotoxins on [^{3}H]TBPS binding to rat brain, avermectin B_{1a}, or the entomopathogenic fungus toxin bassianolide (Abalis, 1981), and the tremorgenic mycotoxin aflatrem (Cole, 1981). The three inhibited [^{35}S]TBPS binding to rat brain membranes (Fig. 11) with IC_{50} of 0.5 μM for avermectin, 2 μM for aflatrem and 7 μM for bassianolide. Avermectin B_{1a} enhanced binding of [^{3}H]Flu as previously reported (Williams and Yarbrough, 1979; Pong et al., 1982), but inhibited [^{3}H]muscimol binding (Fig. 12), which contrasts with a previous report on its potentiation of [^{3}H]GABA binding (Pond and Wang, 1982). The data suggest that these mycotoxins interact with a GABA receptor, and affect its channel binding.

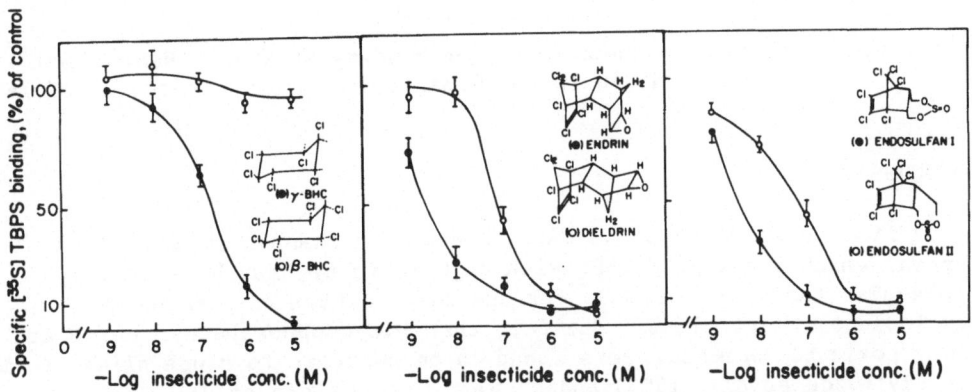

Fig. 10.   Inhibition of the specific binding of [^{35}S]TBPS (2 nM) to
           GABA receptors in rat brain membranes by two cyclodiene and
           one BHC stereoisomeric pairs. All insecticide solutions and
           [^{35}S]TBPS were in ethanol and were added in 10 μl aloquots
           to the incubation tubes. Incubation conditions and time
           were as described in Fig. 8 (from Abalis et al., in press).

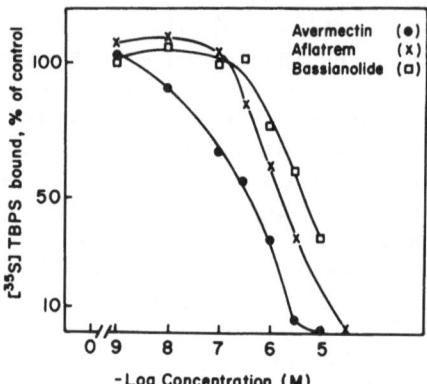

Fig. 11. Effect of three mycotoxins on 2 nM [^{35}S]TBPS binding to rat
brain membranes. Each symbol is the mean of three experiments with SD <10%.

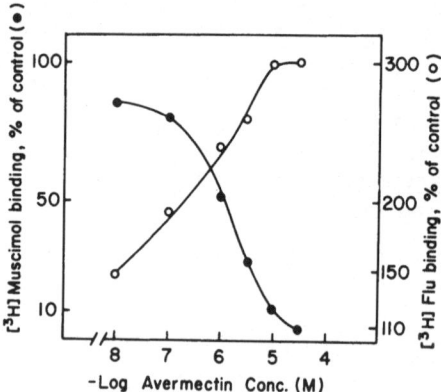

Fig. 12. Effect of avermectin B_{1a} on binding of 4 nM [^{3}H]muscimol
and 1 nm [^{3}H]Flu to the GABA/benzodiazepine receptor of rat
brain.

DISCUSSION

It is evident that the GABA$_A$ receptor's channel in mammalian
brain, which binds [^{35}S]TBPS with high affinity, may be a primary
molecular target in mammals for the toxic action of lindane and
cyclodiene insecticides. The fact that bicyclophosphates, which are
very toxic to mammals, were found to be nontoxic to house flies at 15
μg/fly (Ozoe et al., 1983) and so was picrotoxinin at 100 μg/fly
(Kuwano et al., 1980), and that so far there has been no report of
specific [^{35}S]TBPS binding to insect tissues, may be indicative that
the insect GABA receptors may have lower affinity for this ligand;
thus their channels have different drug specificities. However, the
cyclodienes are potent insecticides possibly as a result of
inhibiting an insect GABA receptor channel. It is hoped that either

receptors, possibly by modifying the membrane preparation or binding assays used, or a higher affinity ligand than [^{35}S]TPBS or [^{3}H]DPTX would be found for the channel site of the insect GABA receptor so as to permit its study. The possible presynaptic action of dieldrin at cholinergic synapses in the sixth abdominal ganglion of the cockroach (Shankland and Schroeder, 1973) may still be due to its inhibition of a GABA$_A$ receptor that in insects may be present also presynaptically on cholinergic neurons. One point is certain, which is that the

GABA/benzodiazepine receptor channel complex carries several kinds of sites that can be potential targets for different kinds of insecticides.

ACKNOWLEDGEMENT

Our reported research was financed in part by NIH grant ES 02594. We are grateful to Dr. R. Cole of the Peanut Institute for providing us with aflatrem, Merck, Sharp & Dohme for avermectin B$_{1a}$, and the Environmental Protection Agency for cyclodienes and BHC. We thank our secretary Ms. Evelyn Elizabeth for her excellent typing.

REFERENCES

Abalis, I.M., 1981, Biochemical and pharmacological studies of the insecticidal cyclodepsipeptides Destruxins and a Bassianolide produced by entomopathogenic fungi, Ph.D. Thesis, Cornell Univ., Ithaca, N.Y.

Abalis, I.M., Eldefrawi, M.E., and Eldefrawi, A.T., 1983, Biochemical identification of putative GABA/benzodiazepine receptors in house fly thorax muscles, Pestic. Biochem. Physiol., 20:39-48.

Abalis, I.M., Eldefrawi, M.E., and Eldefrawi, A.T., High affinity stereospecific binding of cyclodiene insecticides and γ-BHC to GABA receptors of rat brain, Pestic. Biochem. Physiol., (in press)

Barker, J.L., and Mathers, D.A., 1981, GABA analogues activate channels of different duration on cultured mouse spinal neurons, Science, 212:358-361.

Bowery, N.G., Doble, A., Hill, D.R., Hudson, A.L., Shaw, J.A., Turnbull, M.J., and Warrington, R., 1981, Bicuculline-insensitive GABA receptors on peripheral autonomic nerve terminals, Eur. J. Pharmacol., 71:53-70.

Bowery, N.G., Hill, D.R., and Hudson, A.L., 1983, Characteristics of GABA$_B$ receptor binding sites on rat whole brain synaptic membranes, Br. J. Pharmacol., 78:191-206.

Braestrup, C., Nielsen, M., Skovbjerg, H., and Gredal, O., 1981, β-Carboline-3-carboxylates and benzodiazepine receptors, in: "GABA and Benzodiazepine Receptors," E. Costa, ed., Raven Press, New York.

Braestrup, C., and Squires, R.F., 1977, Specific benzodiazepine

receptors in rat brain characterized by high affinity [^{3}H]-
diazepam binding, Proc. Natl. Acad. Sci. USA, 74:3805-3809.

Braestrup, C., and Squires, R.F., 1978, Brain specific benzodiazepine
receptors, Br. J. Psychiatry, 133:249-260.

Casida, J.E., Eto, M., Moscioni, A.D., Engel, J.L., Milbrath, D.S.,
and Verkade, J.G., 1976, Structure-toxicity relationships of
2,6,7-trioxabicyclo(2.2.2)-octanes and related compounds,
Toxicol. Appl. Phrmacol., 36:261-279.

Clifford, D.P., and Jeffrey, P., 1977, The insecticidal and acaricidal properties of some 3-alkylcarbamoyloximino-2,4-dimethyl-1,5-
benzodiazepines, Pestic. Sci., 8:446-448.

Cole, R.J., 1981, Tremorgenic mycotoxins: an update, in: "Anti-
nutrients and Natural Toxicants in Foods," R.L. Ory, ed., Food &
Nutrition Press, Westport, CT.

Cull-Candy, S.G., 1982, Properties of postsynaptic channels activated
by glutamate and GABA in locust muscle fibres, in: "Neuropharmacology of Insects," Ciba Found. Symp. 88., Pitman, London.

Curtis, D.R., Game, C.J.A., and Lodge, D., 1976, The in vivo inactivation of GABA and other inhibitory amino acids in the cat
nervous system, Exp. Brain Res., 25:413-428.

Eldefrawi, A.T., Shaker, N., and Eldefrawi, M.E., 1982, Binding of
acetylcholine receptor/channel probe to housefly head membranes,
in: "Neuropharmacology of Insects," Ciba Found. Symp. 88,
Pitman, London.

Enna, S.J., 1983, GABA receptors, in: "The GABA Receptors," S.J.
Enna, ed., The Humana Press, Clifton, N.J.

Enna, S.J., and Gallagher, J.P., 1983, Biochemical and electrophysiological characteristics of mammalian GABA receptors, Int.
Rev. Neurobiol., 24:181-212.

Fritz, L.C., Wang, C.C., and Gorio, A., 1979, Avermectin B$_{1a}$
irreversibly blocks postsynaptic potentials at the lobster
neuromuscular junction by reducing muscle membrane resistance,
Proc. Natl. Acad. Sci. USA, 76:2062-2066.

Gallagher, J.P., Nakamura, J., and Shinnick-Gallagher, P., 1983,
Effects of glial uptake and desensitization in the activity of
γ-aminobutyric acid (GABA) and its analogs at the cat dorsal root
ganglion, J. Pharm. Exp. Ther., 226:876-884.

Gavish, M., and Snyder, S.H., 1980, Soluble benzodiazepine receptors:
GABA-ergic regulation, Life Sci., 26:579-582.

Ghiasuddin, S.M., and Matsumura, F., 1982, Inhibition of gamma-
aminobutyric acid (GABA)-induced chloride uptake by gamma-BHC
and heptachlor epoxide, Comp. Biochem. Physiol., 73C:141-144.

Kadous, A.A., Ghiasuddin, S.M., Matsumura, F., Scott, J.G., and
Tanaka, K., 1983, Difference in the picrotoxinin receptor
between the cyclodiene-resistant and susceptible strains of the
German cockroach, Pestic. Biochem. Physiol., 19:157-166.

Karobath, M., Drexler, G., and Supavilai, P., 1981, Modulation by
picrotoxin and IPTBO of ^{3}H-flunitrazepam binding to the
GABA/benzodiazepine receptor complex of rat cerebellum, Life
Sci., 28:307-313.

Kuwano, E., Oshima, K., and Eto, M., 1980, Synthesis and insecticidal activity of 8-isopropyl-6-oxabicyclo[3.2.1]octan-7-one, a partial skeleton of picrotoxinin, and related compounds, Agric. Biol. Chem., 44:383-386.

Lawrence, L.J., and Casida, J.E., 1983. Stereospecific action of pyrethroid insecticides on the γ-aminobutyric acid receptor-ionophore complex, Science, 221:1399-1401.

Lawrence, L.J., and Casida, J.E., 1984, Interactions of lindane, toxaphene and cyclodienes with brain-specific t-butylbicyclophosphorothionate receptor, Life Sci., 35:171-178.

Leeb-Lundberg, F., Snowman, A., and Olsen, R.W., 1981, Perturbation of benzodiazepine receptor binding by pyrazolpyridines involves picrotoxinin/barbiturate receptor sites, J. Neurosci., 1:471-477.

Majewska, M.D., and Chuang, D.-M., 1984, Modulation by calcium of γ-amino-butyric acid (GABA) binding to GABA$_A$ and GABA$_B$ recognition sites in rat brain, Mol. Pharmacol., 25:352-359.

Massotti, M., Guidotti, A., and Costa, E., 1981, Characterization of benzodiazepine and γ-aminobutyric recognition sites and their endogenous modulators, J. Neurosci., 1:409-418.

Matsumura, F., and Ghiasuddin, S.M., 1983, Evidence for similarities between cyclodiene type insecticides and picrotoxinin in their action mechanisms, J. Environ. Sci. Health B18, 1-14.

McBurney, R.N., and Barker, J.L., 1978, GABA-induced conductance fluctuations in cultured spinal neurons, Nature, 274:594-497.

Milbrath, D.S., Engel, J.L., Verkade, J.G., and Casida, J.E., 1979, Structure-toxicity relationships of 1-substituted-4-alkyl-2,6,7-trioxabicyclo [2.2.2]octanes, Tox. Appl. Pharmacol., 47:287-293.

Miller, T.A., Mynard, M., and Kennedy, J.M., 1979, Structure and insecticidal activity of picrotoxinin analogs, Pestic. Biochem. Physiol., 10:128-136.

Nielsen, M., Braestrup, C., and Squires, R.F., 1978, Evidence for a late evolutionary appearance of brain-specific benzodiazepine receptors: an investigation of 18 vertebrate and 5 invertebrate species, Brain Res., 141:342-346.

Olsen, R.W., 1983, Biochemical properties of GABA receptors, in: "The GABA Receptors," S.J. Enna, ed., The Humana Press, Clifton, N.J.

Olsen, R.W., Ticku, M.K., and Miller, T., 1978, Dihydropicrotoxinin binding to crayfish muscle sites possibly related to γ-aminobutyric acid receptor-ionophores, Mol. Pharmacol., 14:381-390.

Ong, J., and Kerr, D.I.B., 1983, GABA$_A$- and GABA$_B$-receptor-mediated modification of intestinal motility, Eur. J. Pharmacol., 86:9-17.

Onodera, K., and Takeuchi, A., 1975, Ionic mechanism of the excitatory synaptic membrane of the crayfish neuromuscular junction, J. Physiol., 252:295-318.

Ostlind, D.A., Cifelli, S., and Lang, R., 1979, Insecticidal activity of the anti-parasitic avermectins, Veterinary Record, 105:168.

Ozoe, Y., Mochida, K., Nakamura, T., Shimizu, A., and Eto, M. 1983, Toxicity of bicyclic phosphage GABA antagonists to the housefly Musca domestica L., J. Pesticide Sci., 8:601-605.

Pong, S.-S., Dehaven, R., and Wang, C.C., 1981. Stimulation of benzodiazepine binding to rat brain membranes and solubilized receptor complex by avermectin B_{1a} and GABA, Biochem. Biophys. Acta., 646:143-150.

Pong, S.-S, Dehaven, R., and Wang, C.C., 1982, A comparative study of avermectin B_{1a} and other modulators of the γ-aminobutyric acid receptor chloride ion channel complex, J. Neurosci., 2:966-971.

Pong, S.-S., and Wang, C.C., 1982, Avermectin B_{1a} modulation of γ-aminobutyric acid receptors in rat brain membranes, J. Neurochem., 38:375-379.

Shankland, D.L., and Schroeder, M.E., 1973, Pharmacological evidence for a discrete neurotoxic action of dieldrin (HEOD) in the American cockroach, Periplaneta americana (L.), Pestic. Biochem. Physiol., 3:77-85.

Speth, R.C., Wastak, G.J., Johnson, P.C., and Yamamura, H.I., 1978, Benzodiazepine binding in human brain: characterization using [^{3}H]flunitrazepam, Life Sci., 22:859-866.

Squires, R.F., Casida, J.E., Richardson, M., and Saederup, E., 1983, [^{35}S]t-butylbicyclophosphorothionate binds with high affinity to brain-specific sites coupled to γ-aminocutyric acid-A and ion recognition sites, Mol. Pharmacol., 23:326-336.

Steiner, F.A., and Felix, D., 1976, Antagonistic effects of GABA and benzodiazepines on vestibular and cerebellar neurons, Nature, 260:346-347.

Study, R.E. and Barker, J.L., 1981, Diazepam and (-)-pentobarbital: fluctuation analysis reveals different mechanisms for potentiation of γ-aminobutyric acid responses in cultured central neurons, Proc. Natl. Acad. Sci USA, 78:7180-7184.

Supavilai, P., and Krobath, M., 1981, In vitro modulation by avermectin B_{1a} of the GABA/benzodiazepine receptor complex of rat cerebellum, J. Neurochem., 36:798-803.

Taft, W.C., and Delorenzo, R.J., 1984, Micromolar-affinity benzodiazepine receptors regulate voltage-sensitive calcium channels in nerve terminal preparations, Proc. Natl. Acad. Sci. USA., 81:3118-3122.

Ticku, M.K., Ban, M., and Olsen, R.W., 1978, Binding of [^{3}H] - dihydropicrotoxinin, a γ-aminobutyric acid synaptic antagonist, to rat brain membranes, Mol. Pharmacol., 14:391-402.

Ticku, M.K. and Maksay, G., 1983, Convulsant/depressant site of action at the allosteric benzodiazepine-GABA receptor-ionophore complex, Life Sci., 33:2363-2375.

Ticku, M., and Olsen, R.W., 1979, Convulsants inhibit picrotoxinin binding, Neuropharmacol., 18:315-318.

Van Harreveld, A., 1936, A physiological solution for freshwater crustaceans, Proc. Soc. Exp. Biol. Med., 34:428-432.

Williams, M. and Yarbrough, G.G., 1979, Enhancement of in vitro binding and some of the pharmacological properties of diazepam by a novel anthelmintic agent, avermectin B_{1a}, Eur. J. Pharmacol., 56:273-276.

THE EFFECT OF AVERMECTINS ON INVERTEBRATE GABA NERVOUS SYSTEMS

C.C. Wang

University of California
San Francisco, CA 94143

INTRODUCTION

The avermectins are a family of macrolides produced by cultures of Streptomyces avermilitis (Burg et al., 1979). They consist of eight closely related but different chemical structures separable by reverse-phase high pressure liquid chromatography (Campbell et al., 1983). The structures of the avermectins, largely determined by high resolution mass spectrometry, ^{13}C and ^{1}H nuclear magnetic resonance spectroscopy (Albers-Schonberg et al., 1981) and single crystal X-ray crystallography (Springer et al., 1981), are indicated in Figure 1. Avermectin B_{1a} is the major component constituting 85% of all the avermectins in fermentated cultures. Hydrogenation of a mixture of avermectins B_{1a} and B_{1b} with Wilkinson's catalyst in benzene or toluene at $25°C$ under 1 atmosphere of hydrogen for 20 hours generates a mixture of 22,23-dihydroavermectin B_{1a} (180%) and B_{1b} (20%) which has been assigned the nonproprietary name ivermectin.

The avermectins and ivermectin are all highly active against a variety of invertebrates. Two major phyla, the Nemathelminthes and the Arthropods, are among the most susceptible to them. A broad spectrum of parasitic nematodes in a variety of domestic animals as well as in humans can be readily eliminated by the avermectins at very low dosages. Among the most sensitive parasites are the immature Dirofilaria immitis in dogs (Anderson et al., 1982) and Dictyocaulus viviparus in cattle (Egerton et al., 1981). The vast majority of nematodes can be controlled by a single dose of ivermectin at 0.1 to 0.2 mg/kg to the animal host (Armour et al., 1980). Recent clinical studies in West Africa indicate that ivermectin effectively reduces the number of microfilaria of Onchocerca volvulus in the skin of infected patients (Aziz et al., 1982), and thus suggest its potential utility in controlling human onchocerciasis.

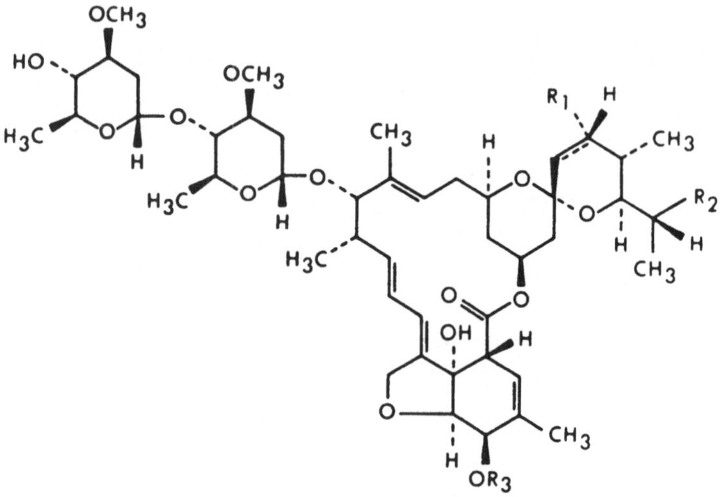

AVERMECTIN	R_1	R_2	R_3
A_{1a}		C_2H_5	CH_3
A_{1b}		CH_3	CH_3
A_{2a}	OH	C_2H_5	CH_3
A_{2b}	OH	CH_3	CH_3
B_{1a}		C_2H_5	H
B_{1b}		CH_3	H
B_{2a}	OH	C_2H_5	H
B_{2b}	OH	CH_3	H

Fig. 1. Molecular structures of the avermectins.

Ivermectin is also highly active against a wide variety of insect (Lancaster et al., 1982) and acarine parasites (Barth and Brokken, 1980). Parasitic fly larvae appear to be particularly susceptible; cattle dosed with 0.2 mg/kg ivermectin indicated full control of all parasitic stages of Hypoderma bovis. The sucking lice Anoplura are more susceptible than the biting lice Mallophaga, whereas Boophilus microplus succumbed to repeated low doses of ivermectin to the cattle and no engorged adult ticks could be recovered. Among the mange mites of cattle, the Psoroptes and Sarcoptes and Sarcoptes scabiei, a single injection of ivermectin to the host at 0.2 mg/kg can achieve a total control.

The avermectins are thus a particularly interesting group of compounds possessing both strong anthelmintic and insecticidal activities. They lack cross-resistance with other commercially available anthelmintics and insecticides, and thus suggest a unique mechanism of action different from all the other known drugs. They are, however, ineffective on species of the phylum Platyhelminthes such as the flukes and tapeworms. They are also inactive against bacteria, fungi, yeast and protozoa. Their toxicity in mammals at the recommended antiparasitic dosage of 0.2 mg/kg is minimal (Campbell et al., 1983). There thus must be specific targets for avermectin action in nematodes and arthropods which is either absent or inaccessible in the microorganisms, flukes and mammals.

In the present report, I would like to trace back to my initial efforts in understanding the mechanism of action of the avermectins, and to describe, in chronological orders, how we eventually overcame the many uncertainties to reach a firm conclusion on this problem. The avermectins act on the invertebrate organisms by potentiating the activity of their GABA (α-aminobutyric acid) nervous systems, which serve the inhibitory function. The avermectin-treated nematodes are immobilized, whereas the drug-treated crustaeceans and insects are paralyzed.

EXPERIMENTAL

Since ivermectin does not represent a single pure chemical entity, avermectin B_{1a} (AVM), the most abundant, active component in the avermectin family, was chosen throughout our investigations. The compound is virtually insoluble in water and is solubilized only in organic solvents such as dimethyl sulfoxide (DMSO), methanol, ethanol, dioxane and chloroform (Miller et al., 1979). In our investigations, AVM solutions made in 100% DMSO were added to the aqueous buffers in various experiments to a final level of 1 to 10% DMSO. Solubilized AVM binds strongly to surfaces of glass and plastics which requires repeated washings with DMSO to be removed. The true concentration of AVM in a solution, which is often below 50% of its calculated value, thus can be estimated only under standardized conditions (the same vessel, solution, volume of solution, temperature, etc.) by a standard curve.

The Negative Findings on Avermectin B_{1a}

Although the avermectins have a common chemical structure resembling that of the macrolide antibiotics, they have no detectable antibiotic activity. AVM exerted no effect on the Escherichia coli ribosome-mediated in vitro protein synthesis or the ^{35}S-methionine incorporation into the trichloroacetic acid-insoluble fraction of the parasitic nematode Dictyocaulus viviparus. It demonstrated no ionophorous activity on the transport of Na^+, K^+, Ca^{++} or Cl^- across erythrocyte membrane, nor had it any inhibitory effect on beef heart mitochondrial ATPase, brain membrane Na^+, K^+-ATPase or Mg^{++}, Ca^{++}-ATPase. Many parasitic nematodes tested by us continued their energy metabolism in a normal state under in vitro incubations; D. viviparus generated lactic acid under aerobic conditions, whereas Trichostrongylus colubriformis carried out CO_2 fixation in its anaerobic metabolism. Neither activity was affected by the presence of up to 100 µg

AVM/ml (Kass et al., 1980). Injection of 1.5 µg of AVM into the
intestinal lumen of a giant parasitic nematode, Ascaris lumbricoides
suum paralyzed the worm, but the worm maintained a normal resting
potential across its muscle membranes (∿−30mV) 24 hours after the
onset of paralysis; suggesting no effect of AVM on energy metabolism
of Ascaris (Kass et al., 1982). AVM also showed no effect on the
assembly of bovine brain microtubules; inhibition of such a process in
nematodes is believed to be the mechanism of anthelmintic actions of
many benzimidazoles (Wang, 1984).

Effects of Avermectin B₁ₐ on the Mobility of Caenorhabditis elegans

The paralytic effect of AVM on A. suum was clearly demonstrated
(Kass et al., 1980). An even more dramatic paralysis of a free-living
soil nematode Caenorhabditis elegans was observed by us (Kass et al.,
1980). These worms were quickly and totally paralyzed upon addition
of AVM to a final concentration of 1 µg/ml. Data in Figure 2
indicated that both the wild-type N2 and the levamisole-resistant
mutant E1072 of C. elegans were equally susceptible to AVM; 50% of the
population was paralyzed by 0.1-0.2 µg/ml of AVM within 10 min. Since
levamisole is known to paralyze nematodes by acting as a nicotinic
acetylcholine agonist at the worm's neuromuscular junctions (Lewis et
al., 1980), the lack of cross-resistance between levamisole and AVM
suggests that the latter does not act on the cholinergic receptor of
C. elegans. This conclusion has been further supported by the total
lack of effect of AVM on a variety of cholinergic systems.

Effects of Avermectin B₁ₐ on Ascaris lumbricoides suum

This giant parasitic nematode has a nervous system very much
similar to that of C. elegans (Stretton et al., 1978). Kass et al.
(1980) first noticed that the Ascaris paralyzed by an injection of AVM
was neither in muscular tetanus nor in flaccid paralysis. Rather, it
retained normal muscular rigidity, showed no significant change in
body-length and has the capability of muscular contraction upon physical stuns. However, pre-injection with AVM did significantly reduce
the shortening of worm-length caused by acetylcholine injection.
Further investigations by Kass et al. (1982) on the Ascaris muscle
strip indicated that AVM did not affect its contraction caused by
applying acetylcholine. But the drug reduced the lengthening of the
acetylcholine-preconditioned muscle-strip caused by GABA. These
findings suggest that AVM may immobilize the nematodes, but it may not
act directly at the neuromuscular junctions of the worm. Although AVM
does not act on the cholinergic receptor, it partially antagonizes the
effect of acetylcholine in Ascaris.

In the Ascaris, the centrally commanding interneurons are in the
ventral cord. There are five repeating arrays of motor nervous
systems each containing 11 neurons of which eight are excitors and
three are inhibitors (Stretton et al., 1978). The former receive

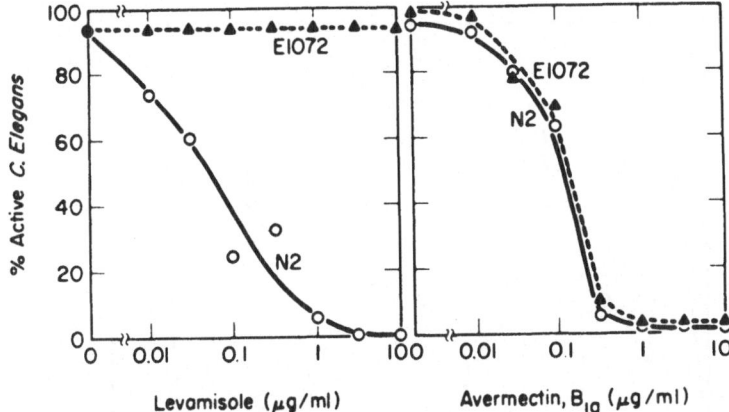

Fig. 2.   Effects of levamisole and AVM on the motility of C. elegans.
Suspension of C. elegans (200–400 adults per ml) were exposed
to drugs at designated concentrations.   After 10 min, the
number of motile worms was counted.   N2, wild-type; E1072,
levamisole-resistant mutant.

input from the interneurons whereas the latter are triggered by the
excitatory motoneurons.   Kass et al., (1980) dissected the anterior
end of adult Ascaris and selectively exposed one pair of intact
commissures of the dorsal excitor #1 (DE1) and the ventral inhibitor
(VI) for electrophysiological studies.   When the DE1 motoneuron was
stimulated directly, AVM exerted no effect on the action potentials
registered from the dorsal muscle cells, thus suggesting no AVM effect
on the DE1-muscle junction.   However, when DE1 was stimulated
indirectly via ventral cord interneurons, the excitatory signals
registered from dorsal muscles were completely abolished by the
presence of AVM (see Figure 3).   This inhibition was, however,
reversed by picrotoxin, a specific blocker of the chloride ion channel
in a variety of neuronal membranes (Takeuchi and Takeuchi, 1969).
Since AVM bound strongly to the membranes, the added picrotoxin could
be readily and selectively washed off, leaving behind the bound AVM
and the re-emergence of its inhibitory effect (Figure 3).   These
observations suggest that AVM acted by blocking signal transmission
from the commanding interneurons to the excitatory motoneurons via
opening the chloride ion channels in neuronal membranes.   Although
little is known about the mechanism of neurotransmission from ventral
cord interneurons to excitatory motoneurons in Ascaris, Kass et al.
(1984) were able to use a divided chamber to indicate that the inter-
neuronal stimulation of DE1 was inhibited by specific application to
the ventral cord of AVM, muscimol, piperazine or curare.   The inhibi-
tions by muscimol and piperazine, two well-known GABA agonists, were
also reversible by picrotoxin.   Thus, AVM blockage of the output of
the VI motoneuron to ventral muscle in Ascaris was also reported (Kass
et al., 1984).   The VI motoneuron has most likely GABA as its neuro-

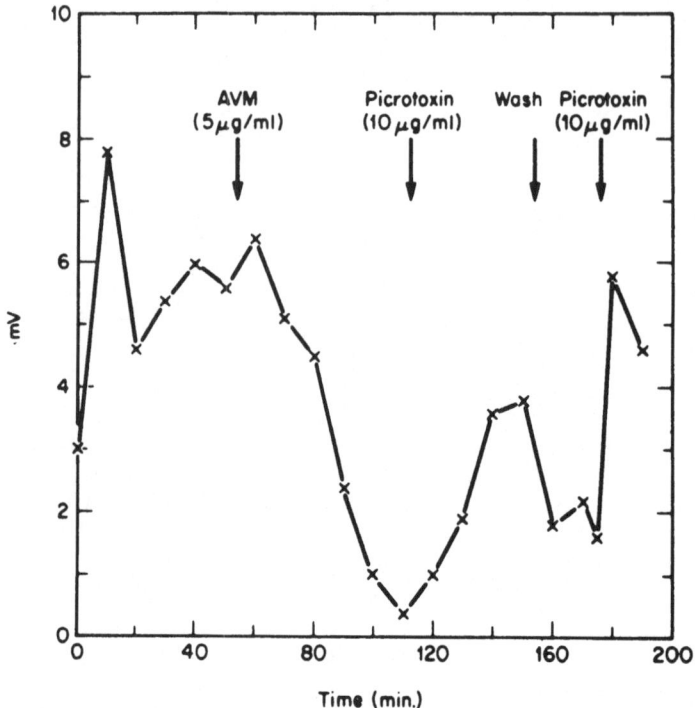

Fig. 3.   Responses to indirect stimulation of dorsal excitatory
          motoneuron #1 (DE1). Each point represents the mean of five
          responses. All recordings are from the same dorsal muscle
          cell. Drugs were added at arrows.

transmitter, which would mean that AVM may have a secondary effect on
Ascaris neuromuscular junction as a GABA antagonist. Picrotoxin,
however, did not block the hyperpolarization induced by VI in concentrations up to 100 µg/ml, which is indicative of chloride ion channels
of distinctive pharmacological properties at Ascaris neuromuscular
junctions. It is possible that the secondary effect of AVM is by
blocking the chloride ion channels at Ascaris neuromucular junctions.

Effects of Avermectin B_{1a} on Daphnia magna

Daphnia magna is a small crustaecean of fresh water readily
immobilized by AVM. At room temperature, about 50% of the D. magna
population was immobilized by 0.1 to 0.2 µg/ml of AVM within 10
minutes (see Figure 4). This potent activity of AVM, comparable to
that observed on C. elegans described in Figure 2, was reversible by
picrotoxin. Data in Figure 4 indicate that the organisms, totally
immobilized by 1.0 ppm of AVM, are almost all brought back to full
motilities with restored heart beats by 100 ppm to 1000 ppm of picrotoxin within 30 minutes. This evidence provides another indication

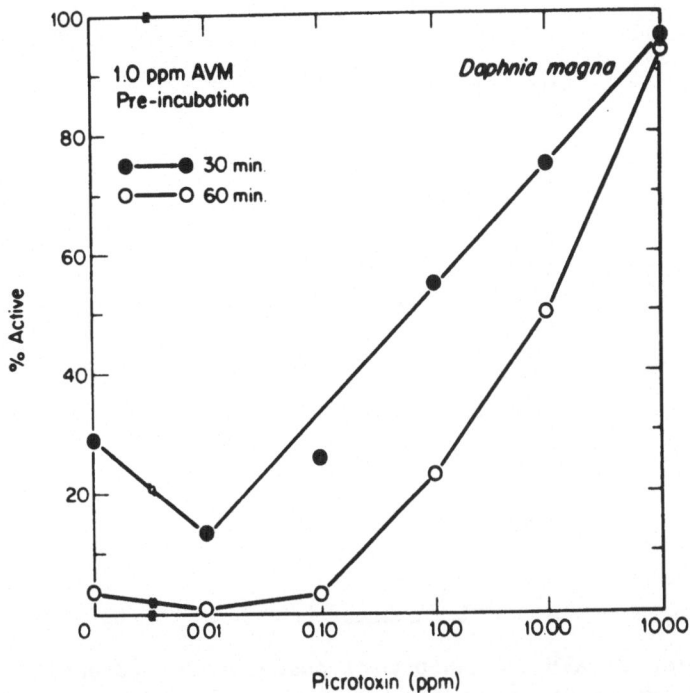

Fig. 4.   Effect of picrotoxin on the motility of D. magna after AVM
treatment at 1.0 μg per ml and room temperature for 30 min
or 60 min. Picrotoxin was then added and incubated as before
for another 30 min.

that the action of AVM on invertebrates is by opening the chloride ion
channels in cellular membranes.

Effects of Avermectin B_{1a} on the Neurotransmission in Lobster's Walking Leg

A model arthropod neuromuscular junction was provided from the
lobster's walking leg by Fritz et al. (1979) for testing possible AVM
effect on signal transmission from nerve to muscle. The stretcher
muscle in a lobster's walking leg is innervated by one excitatory and
one inhibitory axon; the excitatory transmitter is thought to be
glutamate, whereas the inhibitory transmitter is GABA (Gerschenfeld,
1973). Perfusion of the stretcher muscle with 1 to 10 μg/ml of AVM
caused an irreversible elimination of the inhibitory postsynaptic
potentials (IPSPs) within 2 minutes and a more gradual decline of the
amplitude of excitatory postsynaptic potentials (EPSPs) (see Figure
5). The reduction in EPSPs was accompanied by a shortening of EPSP
duration; its phase of repolarization becoming progressively faster
with the duration of AVM treatment. Since the EPSP falling phase is
largely determined by the membrane time constant which is in turn

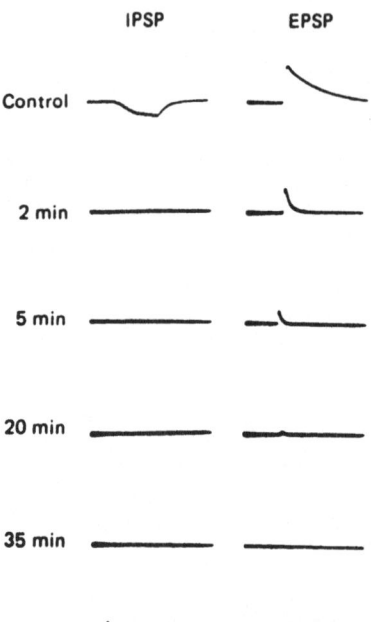

Fig. 5. Effect of AVM on inhibitory postsynaptic potentials (IPSP's)
         and excitatory postsynaptic potentials (EPSP's) in lobster
         muscle. Records were taken 2, 5, 20 and 35 min after the
         application of AVM at 1 µg/ml. Trains of IPSP's (stimulation
         frequency = 30 per sec) were given (left-hand column), and
         single EPSP's were evoked (right-hand column). Calibration:
         IPSP, 0.5 mV, 1 sec; EPSP, 2 mV, 200 msec.

dependent upon membrane resistance (Fatt and Katz, 1951), Fritz et al.
(1979) measured the input resistance in muscle fiber during the drug
application and found that it decreased at the same rate as the EPSP
decline. They also noted that after a few minutes of AVM treatment, a
hyperpolarization of muscle membrane of up to 5 mV was often observed,
which essentially shifted the membrane resting potential toward the
chloride ion equilibrium potential.

    To demonstrate whether the effect of AVM on EPSP's may be
accounted for by its effect on membrane resistance, the intracellular
and extracellular EPSPs from the same muscle fiber were recorded
simultaneously during drug treatment. While the intracellular EPSP
diminished with time as indicated in Figure 5, the extracellular
signal, which is a measure of synaptic current flow independent of
muscle membrane properties, remained unchanged by AVM. Iontophoretically applied glutamate to sensitive muscle fibers elicited

depolarizing intracellular responses; these responses were eliminted by AVM application with the same time course as that of intracellular EPSP. These findings clearly indicate that the presynaptic excitatory signal transmission is not involved in AVM action. AVM must have a postsynaptic action on the muscle membrane. To determine whether AVM caused an augmentation of certain specific ion permeability which may lead to the reduced membrane resistance, Fritz et al. (1979) applied AVM to muscle bathed in K^+-free Ringer solution, which brought the resting muscle membrane potential below the chloride ion equilibrium potential by several millivolts. The IPSP's in such a bathing solution thus become depolarizing. AVM abolished the depolarizing IPSP's just as effectively as its abolishment of hyperpolarizing IPSP's. Apparently, the mechanism of AVM inhibition is by depolarizing the muscle membrane toward the chloride ion equilibrium potential. This evidence strongly suggests that AVM acts by selectively increasing chloride ion permeability across the muscle membrane.

It is well known that GABA receptors regulate the opening of Cl^- channels in crustacean muscle, and that these channels are blocked by picrotoxin. When picrotoxin was applied to the AVM-treated stretcher muscle by Fritz et al. (1979), both the diminished EPSP and reduced input resistance were restored to the near control level, as is indicated in Figure 6. The added picrotoxin was then washed out with lobster saline, and EPSP and input resistance again decreased to low values because of the presence of tissue-bound AVM (see Fig. 6). Later studies indicated that biccucullin, another GABA antagonist known to compete directly with GABA receptor binding, also had the same effect as picrotoxin. These findings support the conclusion that AVM acts on lobster neuromuscular junctions by potentiating a GABA-like action on the postsynaptic GABA receptors which in turn open the chloride ion channels in muscle membranes. The physiological consequence of such an action would be neuromuscular paralysis. Thus, AVM may cause different effects on nematodes and lobster; it blocks interneuron-motoneuron transmission in the nematodes but inhibits motoneuron-muscle transmission in lobster, even though the basic mechanism of both actions may be the same.

Effects of Avermectin B_{1a} on the Stretch Receptor of Crayfish

Similar electrophysiological studies of the effects of AVM on the stretch receptor of crayfish were carried out. The dendrites of a stretch receptor receive inhibitory GABA synapses that are capable of blocking action potential initiations in the cell (Kuffler and Eyzaguirre, 1955). When it was perfused with AVM, the action potentials were blocked so that even a vigorous stretch of the receptor elicited no firing. Picrotoxin reversed the AVM block, in accordance with the suggestion that AVM has an agonistic effect on the GABA receptor-chloride ion ionophore complex (Fritz et al., 1979).

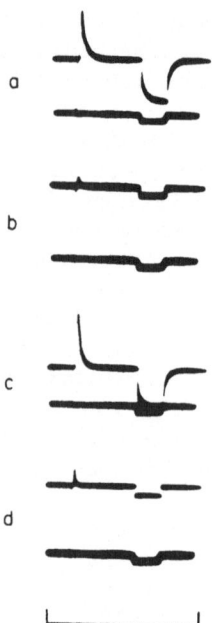

Fig. 6. Effect of picrotoxin on excitatory postsynaptic potential on
         EPSP and on muscle fiber input resistance after AVM treat-
         ment. AVM was applied for 25 min. Picrotoxin was applied
         for 8 min. The preparation was then perfused with lobster
         saline for 27 min. Traces: a. control; b. 24 min after
         application of AVM at 20 μg/ml; c. 2 min after addition of
         picrotoxin at 20 μg/ml; d. 25 min after washing with lobster
         saline. Upper traces record EPSP (left) and response to
         intracellular current pulse (right). Calibration: 2mV, 1
         sec. Lower traces record intracellular current pulse. Cali-
         bration: 100 nA, 1 sec.

DISCUSSION

Our investigations of nematodes and crustaeceans have revealed
that AVM immobilizes all of them. The only agent capable of reversing
this immobilizing effect of AVM is the chloride ion channel blocker
picrotoxin. Similar observations have also been made on the AVM
blockage of signal transmissions from commanding interneurons to motoneurons in Ascaris and from motoneurons to muscle fibers in lobster's
walking legs. The AVM blockade can be specifically removed by picrotoxin, bringing functions of the tissue back to normal. It is thus
apparent that AVM acts on the nematodes and crustaeceans by inducing
the opening of chloride ion channels in the membranes of neurons and
muscle fibers. When the channels in neuronal membranes are selectively opened, such as the case of AVM-treated Ascaris, the animal loses
its central command to motoneurons but retains its normal muscular

contractility. Thus, no flaccid paralysis or muscular contraction is observed in AVM-immobilized Ascaris. In the walking leg of AVM-treated lobster, however, it is the chloride ion channels in the muscle membrane opened by AVM. Thus, AVM-treated arthropods are most likely immobilized by muscular paralysis; a mechanism differing from that underlining AVM immobilization of nematodes. The observed difficulty in washing AVM off the tissue membranes may also explain the high anthelmintic and insecticidal activities of AVM and its derivatives; the accumulation of membrane-bound drug molecules must be increasing steadily with time in treated nematodes or insects. Our previous investigations on C. elegans and D. magna (unpublished) indicated that the IC_{50} value of AVM on the mobility of these organisms is inversely related to the duration of drug treatment. Thus, the IC_{50} value of 0.1 µg/ml AVM on C. elegans recorded after a 10 min incubation (see Figure 2) can be reduced to 0.001 µg/ml when the incubation time is prolonged to two hours. This characteristic drug action has been reflected by the gradual loss of movements among insects treated with very low doses of AVM (Lancaster et al., 1982).

The biochemical mechanism of AVM action has been further pursued by Wang and Pong (1982) using mammalian brain tissues which are known to contain high densities of GABA-receptor-chloride ion channel complex in the synaptic membrane fraction. AVM was found to bind to rat and dog brain synaptosomes with a high affinity; an apparent dissociation constant (Kd) of 1-2 nM and a density of binding sites of 1.54 pmoles/mg protein were estimated (Pong and Wang, 1980). The binding is stereospecific and the affinities of various AVM analogs to the binding site correlate well with their anthelmintic activities. The highest density of AVM binding sites was identified in the cerebellum where the Purkinje cells with GABA nerve endings are most densely populated. The AVM binding, however, does not compete with the bindings of GABA, benzodiazepines or any other naturally occurring neurotransmitters identified in mammalian brains. Our previous data in Figures 3 and 6 showing return of AVM inhibition of neurotransmission after removal of picrotoxin without newly added AVM also pointed to the probability that AVM and picrotoxin do not share the same binding site. There is most likely a specific binding site for AVM associated with the chloride ion channel. AVM was found also causing a significant and long-lasting increase in the rate of GABA release from rat brain synaptosomes (Pong et al., 1980). This stimulation is specific on GABA release from GABA nerve endings, is not by inhibiting re-uptake of GABA by synaptosomes, and is independent of Ca^{++}. This stimulatory effect will certainly contribute to the AVM opening of chloride ion channels. Another AVM effect was observed in its potentiation of postsynaptic GABA binding to rat brain synaptic membranes (Pong and Wang, 1982). The potentiation is by increasing the number of high-affinity GABA receptors in the postsynaptic membrane and is dependent on the presence of chloride ions. Picrotoxin inhibits this AVM effect on GABA receptors, although picrotoxin has no direct interaction with GABA receptor. It is apparent that the AVM potentiation of GABA receptor binding plays a major role in the AVM opening of chloride ion channels.

In conclusion, the mechanism of AVM action has been delineated by us as a combined consequence of enhanced presynaptic GABA release and increased postsynaptic GABA-binding, which brings about the opening of chloride ion channels in the membranes of neurons and muscle fibers. This novel mode of anthelmintic and insecticidal action, never observed before among the previously discovered anthelminthics and insecticides, helps to explain why AVM and its analogs have not demonstrated cross-resistance with other anthelmintics or insecticides (Campbell et al., 1983), known to act as acetylcholinesterase inhibitors, cholinergic agonists or microtubule assembly inhibitors. On the other hand, the antiparasitic therapeutic usefulness of AVM is not due to its lack of similar effects on the GABA-receptor-chloride-ion-channel complex in the mammalian host. But rather, it is due to confinement of GABA nerves in the central nervous system, well protected by the blood brain barrier, in the mammalian host. AVM and its analogs have a rather poor capability of penetrating the blood brain barrier (Campbell et al., 1983) which provides the main basis of its safety margin. When the drug was delivered directly to the rat brain, it acted as an anticonvulsant (unpublished). When AVM was administered at an excessively high level, some drug effect on the mammalian central nervous system could be observed, e.g., when AVM was administered to mice by intraperitoneal injection at the level of 8 mg/kg body weight, it potentiated the muscle relaxing effect of diazepam at suboptimal doses (William and Yarbrough, 1979). This effect was also observed in our in vitro studies, which demonstrated AVM enhancement of benzodiazepine binding to its receptor in the rat brain membranes (Pong et al., 1981). Mydriasis and tremors were seen in dogs dosed orally at 2.0 mg/kg of ivermectin (Campbell et al., 1983); similar effects were much more pronounced in colley dogs which are known to have a deficient blood brain barrier.

Thus, the impressive anthelmintic and insecticidal activities of AVM and analogs are partly due to the ready accessibility of the drugs to the GABA-receptor-chloride ion channels in nematodes and arthropods. This is not only due to the absence of blood brain barrier in these invertebrates but also must be due to ready drug penetration through the cuticles protecting nematodes and insects. It is thus a highly fortuitous discovery of a family of AVM-type compounds which have the observed activities as well as the particular property of only reaching the GABA nerves in the nematodes and insects to make themselves therepautically useful compounds.

REFERENCES

Albers-Schönberg, G., Arison, B.H., Chabala, J.C., Douglas, A.W., Eskola, P., Fisher, M.H., Lusi, A., Mrozik, H., Smith, J.L., and

Tolman, R.L, 1981, Avermectins, structure determinations, J. Am. Chem. Soc., 103:4216-4221.

Anderson, D.L., and Roberson, E.L., 1982, Activity of ivermectin against canine intestinal helminths, Am. J. Vet. Res., 43:1681-1683.

Armour, J., Bairden, K., and Prestron, J.M., 1980, Anthelmintic efficiency of ivermectin against naturally acquired bovine gastrointestinal nematodes, Vet. Rec., 107:226-227.

Aziz, M.A., Diallo, S., Diopp, I.M., Lariviere, M., and Port, M., 1982, Efficacy and tolerance of ivermectin in human onchocerciasis, Lancet, 1982-II, 171-173.

Barth, D., and Brokken, S., 1980, The activity of 22,23-dihydroavermectin B1 against the pig louse, Haematopinus suis, Vet. Rec., 106:388.

Burg, R.W., Miller, B.M., Baker, E.E., Birnbaum, J., Currie, S.A., Hartman, R., Kong, Y.L., Monaghan, R.L., Olson, G., Putter, I., Tunac, J.B., Wallick, H., Stapley, E.O., Oiwa, R., and Omura, S., 1979, Avermectins, new family of potent anthelmintic agents: Producing organism and fermentation, Antimicrob. Agents Chemother., 15:361-367.

Campbell, W.C., Fisher, M.H., Stapley, E.O., Albers-Schönberg, G., and Jacob, T.A., 1983, Ivermectin: A potent new antiparasitic agent, Science, 221:823-828.

Egerton, J.R., Eary, C.H., and Suhayda, D., 1981, The anthelmintic efficacy of ivermectin in experimentally infected cattle, Vet. Parasitol., 8:59-70.

Fatt , P., and Katz, B., 1951, An analysis of the end-plate potential recorded with an intracellular electrode, J. Physiol., London, 115:320-370.

Fritz, L.C., Wang, C.C., and Gorio, A., 1979, Avermectin B_{1a} irreversibly blocks post-synaptic potentials at the lobster neuromuscular junction by reducing muscle membrane resistance, Proc. Natl. Acad. Sci., USA 76:2062-2066.

Gerschenfeld, H.M., 1973, Chemical transmission in invertebrate central nervous systems and neuromuscular junctions, Physiol. Rev., 53:1-119.

Kass, I.S., Larsen, D.A., Wang, C.C., and Stretton, A.O.W., 1982, Ascaris suum: Differential effects of avermectin B_{1a} on the intact animal and neuromuscular strip preparations, Exp. Parasitol., 54:166-174.

Kass, I.S., Stretton, A.O.W., and Wang, C.C., 1984, The effects of avermectin and drugs related to acetylcholine and 4-aminobutyric acid on neurotransmission in Ascaris, Molec. Biochem. Parasitol., 13:213-226.

Kass, I.S., Wang, C.C., Walrond, J.P., and Stretton, A.O.W., 1980, Avermectin B_{1a}, a new paralyzing anthelmintic that affects interneurons and inhibitory motoneurons in Ascaris, Proc. Natl. Acad. Sci., USA, 77:6211-6215.

Kuffler, S.W., and Eyzaguirre, C., 1955, Synaptic inhibition in an isolated nerve cell, J. Gen. Physiol., 39:155-184.

Lancaster, J.L., Simco, J.S., and Kilgore, R.L., 1982, Systematic
    efficacy of ivermectin MK-933 against the lone star tick, J.
    Econ. Entomol., 75:242-244.

Lewis, J.A., Wu, C.H., Levine, J.H., and Berg, H., 1980, Levamisole-
    resistant mutants of the nematode Caenorhabiditis elegans appear
    to lack pharmacological acetylcholine receptors, Neuroscience,
    5:967-989.

Miller, T.W., Chaiet, L., Cole, D.J., Cole, L.J., Flor, J.E.,
    Goegelman, R.T., Gullo, V.P., Joshua, H., Kempf, A.J., Krellwitz,
    W.R., Monaghan, R.L., Ormond, R.E., Wilson, K.E., Albers-
    Schonberg, G., and Putter, I., 1979, Avermectins, new family of
    potent anthelmintic agents: Isolation and chromatographic
    properties, Antimicrob. Agents Chemother., 15:368-371.

Pong, S.S., DeHaven, R., and Wang, C.C., 1981, Stimulation of benzo-
    diazepine binding to rat brain membranes and solubilized receptor
    complex by avermectin B_{1a} and GABA, Biochim. Biophys. Acta,
    646:143-150.

Pong, S.S. and Wang, C.C., 1980, Specific binding of avermectin B_{1a}
    to mammalian brain synaptosomes, Neuropharmacol., 19:311-317.

Pong, S.S., and Wang, C.C., 1982, Avermectin B_{1a} modulation of γ-
    aminoburyric acid receptors in rat brain membranes, J.
    Neurochem., 38:375-379.

Pong, S.S., Wang, C.C., and Fritz, L.C., 1980, Studies on the
    mechanism of action of avermectin B_{1a}: Stimulation of the
    release of γ-aminobutyric acid from nerve endings, J. Neurochem.,
    34:351-358.

Springer, J.P., Arison, B.H., Hirschfield, J.M., and Hoogsteen, K.,
    1981, The absolute stereochemistry and conformation of avermec-
    tin B_{2a} aglycone and avermectin B_{1a}, J. Am. Chem. Soc., 103:4221-
    4224.

Stretton, A.O.W., Fishpool, R.M., Southgate, E., Donmoyer, J.E.,
    Walrond, J.P., Moses, J.E.R., and Kass., I.S., 1978, Structure
    and physiological activity of the motoneurons of the nematode
    Ascaris, Proc. Natl. Acad. Sci., USA, 75:3493-3497.

Tekeuchi, A., and Takeuchi, N., 1969, A study of the action of picro-
    toxin on the inhibitory neuromuscular junction of the crayfish,
    J. Physiol., (London), 205:377-391.

Wang, C.C., 1984, Parasite enzymes as potential targets for antipara-
    sitic chemotherapy, J. Med. Chem. Perspective, 27:1-9.

Wang, C.C., and Pong, S.S., 1982, Actions of avermectin B_{1a} on GABA
    nerves, in: "Membranes and Genetic Disease, Progress in Clinical
    and Biological Research", Sheppard, J.R., Anderson, V.E., and
    Eaton, J.W., eds., Alan R. Liss, Inc., New York. pp. 373-395.

Williams, M., and Yarbrough, G.G., 1979, Enhancement of in vitro
    binding and some of the pharmacological properties of diazepam by
    a novel anthelmintic agent, avermectin B_{1a}, Eur. J. Pharmacol.,
    56:273-278.

THE EFFECTS OF INSECTICIDES ON NEUROSECRETORY PROCESSES IN INSECTS

Ian Orchard

Department of Zoology
University of Toronto
Toronto, Ontario Canada

INTRODUCTION

It is probably true to state that all neurons are secretory, and that any aspect associated with secretion may be specialized in any given neuron. The net result is that there is probably a continuum of secretory activities evident within the nervous system. Only at the extremes is it possible to provide clear definitions and this has led to the recognition of two basic forms of chemical communication. At one extreme are neurons (conventional neurons) which have axons projecting directly to their target site, and which are specialized to secrete precisely metered small amounts of neurotransmitter. Communication is rapid and consists of short messages delivered across specialized contact areas called synapses. At the other extreme are neurons (neurosecretory cells) which are specialized to secrete relatively large amounts of neurohormone into the circulatory system. These neurohormones are capable of regulating multiple and distant target cells, and the end result is a relatively general, persistent message. In between these extremes are neurons which are hard to strictly define. They appear to possess characteristics of both forms of communication; examples include those neurons which project directly to target tissues, do not form synapses, but merely release neurochemicals into the general vicinity of the target cells.

The classical neurosecretory cells of the Scharrers (review by Gabe, 1966), (i.e. those neurons which secrete neurohormone into the circulatory system), are found extensively in invertebrates and vertebrates and distributed widely throughout the nervous system of insects (Fig. 1). Indeed, a survey of hormones found in insects reveals that most insect hormones are neurohormones (review by Orchard and Loughton, 1984). Exceptions to this include ecdysone and juvenile hormone which are products of non-neural tissue, although it should be pointed out that the activities of these tissues are

139

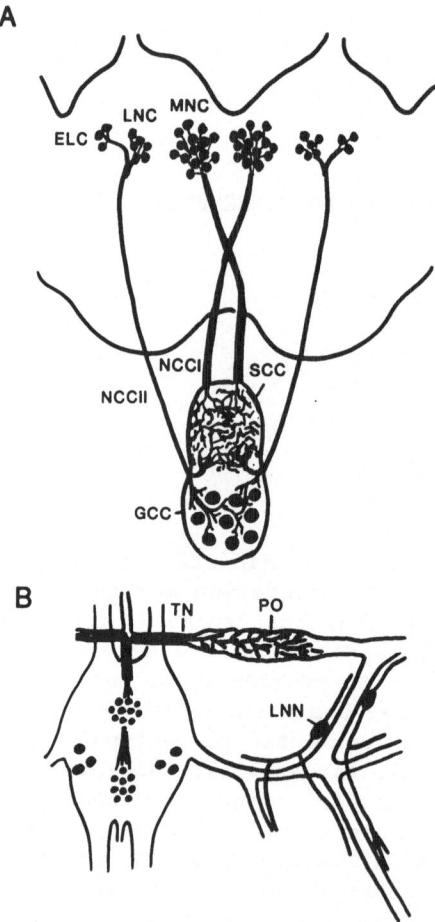

Fig. 1.   Diagrammatic representation of distribution of neurosecretory
          cells (filled circles) and neurohaemal organs in A, brain and
          B, abdominal ganglion of a "typical insect". (A) The corpus
          cardiacum, as represented by the locust, is composed of a
          storage lobe (SCC) and a glandular lobe (GCC). The glandular
          lobe consists of intrinsic neurosecretory cells which are
          synaptically controlled by extra-lateral cells (ELC) in the
          brain which are probably aminergic. These cells project to
          the glandular lobe via the nervi corporis cardiaci II (NCCII).
          The storage lobe is the storage and release site of hormones
          manufactured in the medial neurosecretory cells (MNC) and
          lateral neurosecretory cells (LNC). These cells project to
          the storage lobe via the nervi corporis cardiaci I (NCCI) and
          NCCII respectively. (B) In an abdominal segment, as repre-
          sented by the stick insect, the major neurohaemal organ is the
          perisympathetic organ (PO) lying on the transverse nerve (TN).
          Neurosecretory cells within the central nervous system project
          into the PO. In addition there are neurosecretory cells, the
          link nerve neurosecretory cells (LNN), lying on nerves in the
          periphery. Redrawn from Orchard and Loughton (1984).

probably under the influence of neurohormones. Neurohormones have
been implicated in the control of a plethora of activities, including
developmental, metabolic, physiological and behavioural (reviews by
Maddrell and Nordmann, 1979; Orchard and Loughton, 1984). Any dis-
turbances in the neurosecretory system of insects may thus be anticipated to result in an inbalance of a host of bodily functions.

A variety of insecticides exert their primary action upon the
nervous system of insects, with both nervous conduction and synaptic
transmission being affected. Debate has often centered around the
possibility that some neurons may be more susceptible than others and
in particular attention has focused upon the central versus peripheral nervous system. Over the past few years however there has
been an increasing awareness of the importance of neurosecretory
cells in insects and to the possible involvement of neurosecretory
cells in the eventual poisoning of insects (see Casida and Maddrell,
1971; Orchard, 1980a,b; Normann, 1980; Singh and Orchard, 1982). It
is the purpose of this review to examine the evidence which indicates
neurosecretory cells may be target sites for insecticides. The
review will begin with a description of the specialized features of
neurosecretory cells which may make them more susceptible to insecticides, and continue with a review of the evidence illustrating dis-
turbances in the neurosecretory system following exposure to insecticides.

REVIEW OF RELATED WORK

Insect neurosecretory cells resemble conventional neurons in
gross morphology (Fig. 2). Both types of neurons possess a cell body
from which emanate a variety of processes including dendrites and/or
axons. The neurosecretory cells within the central nervous system
are typically monopolar, those in the peripheral nervous system are
multipolar (see Orchard, 1983).

Ultrastructural studies reveal that a characteristic feature of
neurosecretory cells is the presence of large numbers of electron-
dense granules. These neurosecretory granules apparently originate
within the Golgi apparatus (Normann, 1965; Finlayson and Osborne,
1968; review by Orchard and Loughton, 1984) and are believed to
contain a carrier protein and neurohormone. These granules are
transported to the terminals, which are specialized for storage and
release of neurohormone (Fig. 2). The axons of neurosecretory cells
repeatedly divide into fine branches, each of which terminate in a
swollen ending which may contain large numbers of electron-dense
granules. By this means the surface area over which release may
occur is greatly expanded. These branched terminals must lie outside
the 'blood-brain barrier' to allow the neurohormone easy access to
the haemolymph. Consequently the axon terminals penetrate the
perineurium surrounding the central nervous system, thereby leaving

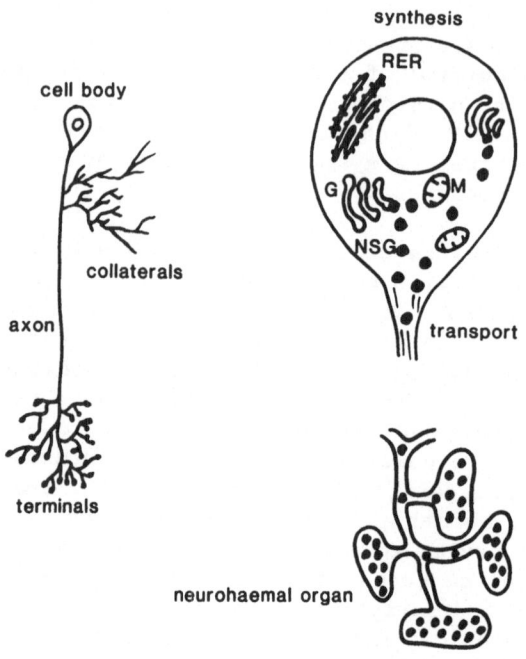

Fig. 2.   Diagrammatic representation of a 'typical' insect neurosecre-
          tory cell found within the central nervous system. The cell
          body is the site of synthesis of neurosecretory material which
          is packaged into membrane-bound neurosecretory granules (NSG)
          by the Golgi apparatus (G).   These NSG are transported along
          the axons to the terminals where they undergo storage and
          release. M, mitochondrion. Redrawn from Orchard and Loughton
          (1984).

their protective covering, and lie naked on the outside of peripheral
nerves (Brady and Maddrell, 1967; Normann, 1965; Finlayson and
Osborne, 1968).   Since many different neurosecretory cells send axons
to the same area for release, the outside of the peripheral nerve
becomes swollen with numerous branched endings, and these swellings
are termed neurohaemal organs (Fig. 1).   The corpus cardiacum is
essentially the neurohaemal organ for neurosecretory cells in the
brain, while the neurosecretory cells in the ventral ganglia have
axons within the median nervous system which terminate in the perisympathetic organs. Diffuse neurohaemal areas are also found extensively along many segmental nerves in insects (eg. Fifield and
Finlayson, 1978). It is important to realize from the above description that there are large areas of membrane exposed to the haemolymph
and thereby exposed to any chemical within the haemolymph.

Neurosecretory cells are specialized neurons, but have maintained all of the electrical properties of conventional neurons

(reviews by Orchard, 1983; Orchard and Loughton, 1984). Certain features may, however be unique to neurosecretory cells. For example, a fairly consistent finding in insect neurosecretory cells is that they have cell bodies which can support overshooting action potentials. Conventional motor neurons or interneurons do not appear to possess this property. In addition the action potentials of neurosecretory cells are of prolonged duration, being some 2-20 times wider in duration than those of conventional neurons, and calcium appears to play a major role as a charge carrier in the rising phase of the action potential. Thus while the general features of the neurophysiological properties of neurosecretory cells and conventional neurons may be similar, there are subtle differences in the manifestation of these features.

It is generally accepted that neurosecretory cells release their neurohormones by the mechanism of exocytosis, a process whereby the neurosecretory granule fuses with the plasma membrane thereby producing a characteristic omega profile (or exocytotic pit). Evidence for this has accumulated from the number of studies in which there has been an increase in omega profiles at times when hormone release was believed to be occurring (review by Orchard, 1984).

The most obvious and clearly established function for the electrical activity of neurosecretory cells is to regulate the release of neurohormone - as in conventional neurons where action potentials result in the release of neurotransmitter. In insects, electrical stimulation of neurosecretory cells has been shown to induce histological and ultrastructural changes indicative of release (Gosbee et al., 1968; Scharrer and Kater, 1969; Krogh and Normann, 1977). More direct evidence has been obtained by bioassaying for hormone release following electrical stimulation (Kater, 1968; Gersch et al., 1970; Orchard et al., 1981). Direct depolarization of neurosecretory terminals using elevated potassium salines has led to hormone release from abdominal neurohaemal organs (Maddrell and Gee, 1974) and from the corpus cardiacum (Orchard and Loughton, 1981b; Orchard et al., 1981). This release is prevented when calcium ions are removed from the high potassium salines. Similarly the electrically elicited release of hormone from the corpus cardiacum is also dependent upon the presence of calcium ions. As a result it is generally accepted that the 'calcium hypothesis' proposed for neurotransmitter release, holds true for neurohormone release.

Many neurosecretory cells exhibit a bursting pattern of electrical activity, and it appears that this greatly increases the amount of hormone released per action potential (see Gainer, 1978; see Orchard, 1983). The correlations of action potentials with neurohormone release suggests that an examination of the patterning and intensity of electrical activity of neurosecretory cells may be used as criteria for determining the timing of hormone release. Studies using this technique (Orchard and Steel, 1980) indicate that this approach may well be a useful one in examining for hormone release.

EXPERIMENTAL

There have been three approaches used to assess the effects of insecticides upon neurosecretory cells. Ultrastructural studies have examined for evidence of increased rates of exocytosis, a sensitive indicator of release of neurohormone. Electrophysiological studies have been performed to examine for a disruption in the normal electrical activity of neurosecretory cells; any increases in electrical activity may be associated with alterations in neurohormonal output. Finally, a number of studies have looked for the release of neurohormone by making use of the various bioassays which indicate the presence of these hormones.

Ultrastructural Studies

Neurohaemal organs are the site of storage and release of neurohormones. These are distinct structures which can be routinely dissected from insects and processed for electron-microscopy. As such therefore it is a relatively simple procedure to expose the neurohaemal organs to insecticides and then observe for effects on their ultrastructure.

Normann and Samaranayaka-Ramasamy (1977) reported the effects of lindane upon the intrinsic neurosecretory cells of the corpus cardiacum of Schistocerca gregaria (see Fig. 1). These neurosecretory cells comprise the glandular lobe of the corpus cardiacum and are the source of adipokinetic hormones (Stone and Mordue, 1979; Carlsen et al., 1979). Exposure of the glands in situ to 5 µl of acetone containing 500 µg lindane/g resulted in ultrastructural evidence for an increase in neurohormonal release. There was an increase in the incidence of exocytotic pits, and the occurence of neurosecretory granules fusing with other granules already in the process of exocytosis. Mitochondria were also affected, with mitochondria in adjacent neurosecretory cells lying juxtaposed, and evidence of mitochondrial division. Singh et al. (1982a) examined the effects of a 20 min in vitro incubation of isolated corpora cardiaca of Locusta with saline containing 1 µM bioresmethrin. Bioresmethrin induced the formation of large vacuoles in the cell bodies of the intrinsic cells of the glandular lobe and exocytotic pits in the axon terminals. There was a depletion of granules in these intrinsic cells, as well as in the extrinsic axons of the corpus cardiacum which arise from cell bodies in the brain. Mitochondria were found to migrate towards the cell membrane and to form small clusters. Pre-incubation with tetrodotoxin blocked the action of bioresmethrin. Similarly permethrin or bioresmethrin in Calliphora (Normann and Samaranayaka-Ramasamy, 1977) and bioresmethrin in <u>Rhodnius</u> (Singh et al., 1982a) increased the secretory rate of axon terminals in the corpus cardiacum as judged by the number of exocytotic pits and the presence of vacuoles. In <u>Rhodnius</u> there was also a dramatic effect upon the mitochondria, which were found clustered in vacuoles, abnormally elongated and again showing signs of fission. Topical application of lindane caused a reduction in the number of neurosecretory granules in the neurohaemal organ on the medial nerve in <u>Carausius morosus</u> (Osborne, 1979) and stimulated the occurence of vacuoles in the intrinsic neurosecretory cells in the glandular lobe of <u>Locusta</u> (Moreteu and Ramade, 1979).

These results all demonstrate quite clearly that exposure of neurohaemal organs to insecticides induced ultrastructural changes indicative of release of neurohormone. The ability of these insecticides to act upon isolated neurohaemal organs <u>in vitro</u> indicates that such changes may be expected to be induced by a direct action upon these structures <u>in vivo</u>.

<u>Electrophysiological Studies</u>

In <u>Carausius morosus</u> the perisympathetic neurohaemal organs lying on the transverse nerve in each segment are electrically active, and action potentials can be recorded from these tissues when the nerves are completely isolated from the rest of the nervous system (Finlayson and Osborne, 1970). Similarly action potentials can be recorded from peripheral neurosecretory cells which lie with their cell bodies on one of the peripheral nerves (the link nerve). These link nerve neurosecretory neurons continue to generate action potentials when the link nerve is isolated (Orchard and Finlayson, 1976). Both of these preparations show a remarkable sensitivity to permethrin (Orchard and Osborne, 1979). Irrigation of isolated preparations of either the link nerve or transverse nerve (Fig. 3) with permethrin caused a massive increase in the frequency of 'spontaneously' generated action potentials. Trains of impulses developed 2-5 min after application of 5×10^{-8} M permethrin. Concentrations as low as 5×10^{-11} M were also effective at altering the spontaneous activity within 7-10 min. Similar effects were obtained with DDT, but much higher concentrations were necessary (5×10^{-6} M). The ability of these insecticides to act upon the isolated neurosecretory systems <u>in vitro</u> indicates again that a direct action may be anticipated <u>in vivo</u>.

In <u>Rhondnius</u> it is possible to record from neurosecretory axons as they enter the corpus cardiacum from the brain (Orchard and Steel, 1980). On-going electical activity can be recorded <u>in situ</u>, in an essentially intact insect, by merely removing a flap of cuticle from the dorsal surface of the head and applying a suction electrode to the exposed corpus cardiacum. Irrigation of the exposed corpus cardiacum with 10^{-6}M of bioresmethrin, bioallethrin, NRDC 161 or permethrin produced large increases in the frequency of action

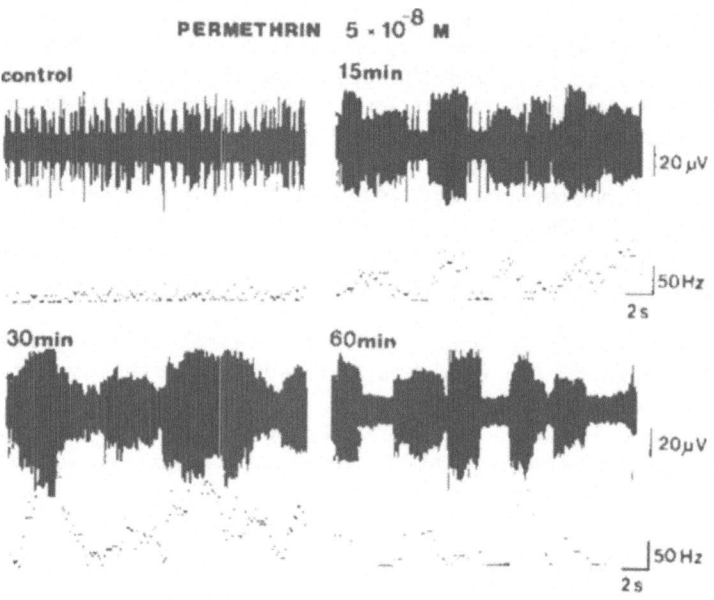

Fig. 3.   Effect of permethrin on 'spontaneous' electrical activity
recorded from the isolated transverse nerve neurohaemal organ
in Carausius morosus. Upper traces represents recordings at
varying times after application of the insecticide. Lower
traces represent frequency of action potentials from the
recordings. From data of Orchard and Osborne (1979).

potentials (Orchard, 1980a,b). This increase in activity was brought
about by a recruitment of units and the production of a phasic
pattern or activity. The response to bioresmethrin and permethrin
reached a peak within 5 min of application, and then maintained a
plateau well above control frequencies. With bioallethrin and
decamethrin the first well defined peak was absent and the increase
in frequency was a gradual one. The relative effectiveness of these
pyrethroids was examined by comparing the concentrations which at 40
min doubled the frequency over that of the control recordings. These
were found to be 10^{-10} M for NRDC 161; 2×10^{-10} M for bioresmethrin;
10^{-9} M for permethrin and 2×10^{-7} M for bioallethrin. To test
whether the neurosecretory cells were activated following poisoning
of the intact animal, an LD 95 of bioresmethrin was topically applied
to unfed male 5th instars and recordings obtained at various times
after poisoning. Electrical activity increased in frequency within
15 min and by 1 hr had reached 300% that of controls. This hyperactivity continued up to the final recording at 24 hr when the insect
was paralyzed (Fig. 4).

The corpus cardiacum of locusts is divided into a storage lobe,
consisting of terminals of neurosecretory cells located in the brain,

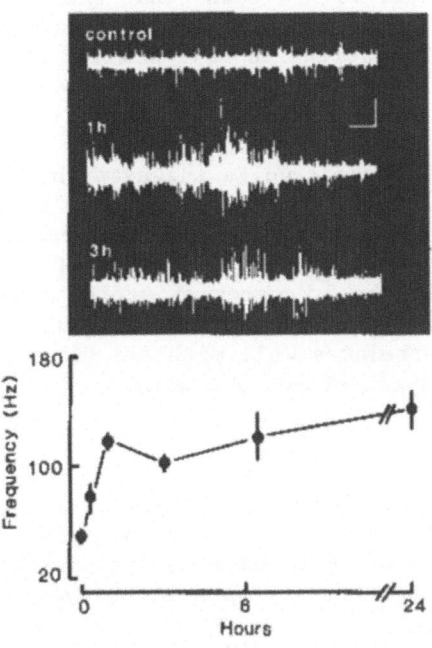

Fig. 4. Topical application of LD 95 bioresmethrin to Rhodnius results in an increase in frequency and change in patterning of electrical activity recorded from the corpus cardiacum in situ. Upper trace represents recordings made from control insect and from insects 1 hour and 3 hours after application of LD 95. Scale bars 100msec; 100 μV. Lower graph illustrates changes in frequency at varying times after topical application of LD 95. From Orchard (1980b).

and a glandular lobe consisting of intrinsic neurosecretory cells. The glandular lobe also contains extrinsic aminergic axons which synaptically control the release of adipokinetic hormones from these cells (see Orchard, 1982). Both the storage lobe and glandular lobe have low levels of spontaneous activity when the corpus cardiacum is isolated in vitro. The source of these action potentials has not been determined but they are probably associated with either neuro-secretory or aminergic cells. A variety of insecticides increase the frequency of spontaneous activity in both the isolated storage and glandular lobes, and induce repetative discharge following the stimulation of a compound action potential via the nerves NCCI and NCCII which supply these lobes (Singh and Orchard, 1982, 1983). For example the frequency of spontaneous action potentials in the storage lobe was increased by application of bioresemthrin (10^{-7} M), DDT, dieldrin, lindane and chlorfenvinphos (10^{-6} M). Bioresmethrin (10^{-7} M), dieldrin and chlorfenvinphos (10^{-5} M) increased the frequency recorded from the glandular lobe. In a comparative study

(Singh and Orchard, 1982) it was found that the insecticides tested modulated the electrical activity of the isolated corpus cardiacum at doses lower than those required to have similar effects on other regions of the central nervous system. Thus the corpus cardiacum appeared more susceptible than the rest of the nervous system.

The evidence from these studies suggests that insecticides alter the electrical activity of neurosecretory cells and that they may have a direct effect upon neurohaemal structures. Since electrical activity is coupled to the process of release of neuro-hormone, alterations in the frequency of action potentials would be expected to induce alterations in hormone output. Thus the increased electrical activity correlates well with the ultrastructural observations of increased release of neurohormone.

Bioassay

The ultimate test of whether insecticides induce hormone release is of course to detect the presence of hormone in poisoned insects. Unfortunately there are no radioimmunoassays available for insect neurohormones and one has to rely upon bioassay.

There is now considerable evidence for insecticide-induced release of neurohormone in insects. Early studies indicated that toxins, other than the insecticides, were liberated into the haemo-lymph of insecticide-poisoned insects (Sternberg et al., 1957; Colhoun, 1958). These toxins, when injected into control insects produced poisoning symptoms similar to those caused by insecticide treatment. The toxic substances reported by these investigators were found to originate either from the nerve cord (Sternberg et al., 1957) or from the corpora cardiaca (Colhoun, 1959). With the heightened awareness of the importance of neurohormones in insects, several studies have looked for an increased presence of neurohor-mones following insecticide poisoning. These studies have either bioassayed material released from isolated neurohaemal organs treated in vitro with insecticide (Singh and Orchard, 1982, 1983; Singh et al., 1982b); assayed the haemolymph for the presence of neurohormone (Casida and Maddrell, 1971); or examined for an effect upon tissue known to be regulated by neurohormone (Singh and Orchard, 1982, 1983; Casida and Maddrell, 1971; Maddrell and Reynold, 1972; Samaranayaka 1976, 1977).

In Rhodnius, paralysis of fifth instar larvae caused by treat-ment with chlorinated hydrocarbon, organophosphorous, methylcar-bonate, pyrethroid and other types of insecticide, resulted in a rapid secretion by their Malpighian tubules (Casida and Maddrell, 1971). This action was not a direct one upon the tubules but involved the release of diuretic hormone from neurosecretory cells in the mesothoracic ganglionic mass. Bioassay of the haemolymph from insecticide treated Rhodnius indicated the increased presence of diuretic hormone activity at the time of paralysis. Nerve section experiments indicated that the insecticides were probably acting via the central nervous system to activate the neurosecretory cells. Similarly, representatives of each of the major classes of insecticides were found to induce plasticization of the abdominal cuticle, an effect produced by a plasticizing factor released from nerve endings in the epidermis (Maddrell and Reynolds, 1972).

Injection of a variety of insecticides into adult female blowflies resulted in alterations in the haemolymph trehalose concentrations (Normann, 1980). Thirty minutes after injection of lindane, haemolymph trehalose levels were decreased by 57%. This effect was dependent upon the presence of the medial neurosecretory cells in the brain, and the presence of intact nervous connections to the corpus cardiacum. This indicates that the effect was due to the release of hypotrehalosemic hormone found within these cells. Pyrethroids increased trehalose levels, and were still effective when the nervous connections to the corpus cardiacum were severed. The pyrethroids were ineffective, however, following extirpation of the corpus cardiacum, indicating that the pyrethroids were acting on the hypertrehalosemic neurosecretory cells found within the corpus cardiacum. In contrast to the effects of lindane in blowflies, Orr and Downer (1983) found that lindane produced effects upon carbohydrate metabolism in cockroaches which resulted from lindane-induced release of aminergic modulators.

More extensive studies have been performed upon the corpus cardiacum of locusts in which lies a convenient system for the detailed study of insecticide-induced release of hormone. The glandular lobe of the corpus cardiacum is the source of two peptidergic adipokinetic hormones (Stone and Mordue, 1979; Carlsen et al., 1979). The release of adipokinetic hormones from the intrinsic neurosecretory cells of the glandular lobe has recently been characterized (Orchard and Loughton, 1981a,b) and shown to be under the immediate control of axons in NCCII which pass from the brain to form synapses with the cells. The transmitter is aminergic, probably octopamine (Orchard et al., 1983a,b). A second input, apparently modulating the output of this synapse, originates from NCCI but is not fully characterized (Orchard and Loughton, 1981a).

In a number of articles, Samaranayaka (1974, 1976, 1977, 1978) examined the effects of insecticide poisoning on the release of adipokinetic hormones in locusts. Topical application of the anticholinesterases, Baythion and Zectran, and the pyrethroid NRDC 119 to Schistocerca resulted in an elevation in haemolymph lipid (Samaranayaka, 1976). Reserpine treatment, which depletes aminergic stores, prevented the elevation induced by the anticholinesterases but not the elevation induced by the pyrethroid. Similarly, injection of aminergic antagonists inhibited the actions of the anticholinesterases, but not the pyrethroids. The conclusions from these studies was that the site of action of the anticholinesterases lay essentially within the brain, and that hormone release was induced by way of cholinergic input activating the aminergic axons which control the neurosecretory cells in the glandular lobe (Samaranayaka, 1976, 1977). The pyrethroid was capable of bypassing this route and presumably acted directly upon the corpus cardiacum. Singh and Orchard (1982, 1983) and Singh et al. (1982b) confirmed and extended these observations. Injection of a variety of organo-chlorine, organophosphorous and pyrethroid insecticides into Locusta, resulted in an elevation in haemolymph lipid (Table 1).

Injected insecticides did not affect the haemolymph lipid levels of neck-ligated locusts, indicating that the elevation in haemolymph lipid in intact locusts was probably due to hormone released from the head.

Incubation of isolated corpora cardiaca in vitro in saline containing the same insecticides induced the release of adipokinetic hormones as judged by bioassay of the medium (Table 2). Among the organochlorine insecticides tested at 10^{-6} M, DDT was most effective, while dyfonate was the most potent organophosphorous insecticide.

Table 1. Insecticide-induced Release of Hyperlipemic Hormone in Locusts in vivo[a]

Insecticide injected	Haemolymph concentration (μM)	Haemolymph lipid (μg/μl)[b]	Percentage change compared with control
DDT	1	4.1 ± 0.05	10 (NS)[c]
	10	13.4 ± 0.4	252 (P<0.005)[d]
Lindane	1	5.7 ± 0.8	50 (NS)
	10	19.9 ± 2.0	424 (P<0.005)
Dieldrin (HEOD)	1	4.5 ± 0.4	18 (NS)
	10	9.2 ± 0.3	142 (P<0.005)
Chlorfenvinphos	1	3.9 ± 0.6	3 (NS)
	10	12.4 ± 0.6	226 (P<0.005)
Bioresmethrin	0.1	5.0 ± 1.0	19 (NS)
	1	24.2 ± 1.8	476 (P<0.005)

[a]Data represent lipid content of locusts injected with two doses of each insecticide. Lipid levels determined 1 hr after injection.
[b]Mean ± SE of four locusts.
[c]NS, not significant.
[d]Figures given in parentheses indicate the level of significance (Student's t test) compared to control injections.
From Singh and Orchard (1982, 1983).

Table 2. Effect of Insecticides on the Release of Hyperlipemic Hormone from Glandular Lobe of Isolated CC of Locusts[a]

Insecticide	Haemolymph lipid (μg/μl)[b]		Percentage increase in haemolymph lipid
	Control	Treated	
DDT	4.5 ± 0.8	10.5 ± 1.2	133 (P< 0.05)[c]
TDE	4.6 ± 0.3	8.2 ± 0.5	78 (P< 0.05)
Lindane	4.7 ± 0.2	6.5 ± 0.5	38 (P = 0.05)
Heptachlor	3.9 ± 0.3	5.7 ± 0.2	46 (P< 0.05)
Dieldrin (HEOD)	4.1 ± 0.3	6.5 ± 0.2	58 (P< 0.05)
Dieldrin-pentachloroketone	4.6 ± 0.5	7.6 ± 0.2	65 (P< 0.05)
Malathion	3.3 ± 0.4	6.1 ± 0.5	85 (P< 0.05)
Chlorfenvinphos	5.1 ± 0.3	9.0 ± 1.0	76 (P< 0.05)
Dyfonate	3.9 ± 0.4	8.2 ± 0.6	110 (P< 0.01)
Bioresmethrin	4.7 ± 0.3	12.2 ± 0.7	160 (P< 0.025)

[a]Data show results of hyperlipemic activity.
[b]Mean ± SE of four locusts.
[c]Figures given in parentheses indicate the level of signifcance (Student's t test).
[d]Bioresmethrin tested at 0.1 μM, all others at 1 μM.
From Singh and Orchard (1982, 1983).

Bioresmethrin was effective at 10^{-7} M.

In a pharmacological study on the effects of insecticides on corpora cardiaca in vitro (Singh et al., 1982b) it was found that reserpine treatment, or application of α-aminergic antagonists had no effect upon the action of DDT or bioresmethrin, but partially blocked the action of dieldrin and chlorfenvinphos. The cholinergic blocker hexamethonium bromide abolished the effects of dieldrin and chlorfenvinphos but again did not affect DDT or bioresmethrin-induced release of hormone. The action of all of these insecticides however was completely blocked by treatment with 10^{-6} M tetrodotoxin. The conclusion from this work is that DDT and bioresmethrin probably act directly upon the intrinsic neurosecretory cells of the glandular lobe, via a sodium-dependent mechanism. Dieldrin and chlorfenvinphos each appear to act via two distinct cholinergic pathways. One of these pathways acts via the pre-synaptic aminergic terminals which control the intrinsic cells, whereas the other acts elsewhere in the corpus cardiacum. Sodium channels are also involved in the ultimate expression of these two insecticides as well.

The results of Singh et al. (1982b) are essentially in agreement with those of Samaranayake (1976) in that she considered the actions of Baythion, an organophosphate and Zectran, a carbamate,

were via cholinergic receptors. The anticholinesterase-induced lipid
elevation in topically poisoned locusts was reduced by aminergic
antagonists. However, her results clearly show that the effects were
not abolished, but merely reduced. Similarly when smaller doses of
reserpine were applied over a prolonged period of time, the Baythion-
induced release of hormone was reduced, but not abolished. Clearly,
in the intact locust Samaranayaka (1976) demonstrated that amines
were necessary for inducing in part, the release of hormone by
anticholinesterases, but her results clearly show that the anticholinesterases were, in addition, acting via at least one other
pathway. This was also indicated by the effectiveness of acetylcholine in releasing hormone from isolated corpora cardiaca
(Samaranayaka, 1977).

DISCUSSION

    The process of toxic action in insects is obviously complex.
An initial action upon a primary site would obviously result in
secondary and tertiary effects. Hyperactivity of the nervous system
has generally been believed to be the result of the action of
insecticides upon their primary site. Neurosecretory cells are under
the integrative control of the nervous system and may be activated
via hyperactivity of the neuronal pathways which impinge upon them.
This appears to be the case in Rhodnius for release of diuretic
hormone (Cassida and Maddrell, 1971). However in other neurosecretory systems (Orchard and Osborne, 1979; Singh and Orchard, 1982,
1983; Normann, 1980), the neurosecretory cells appear to be directly
affected by the insecticides, and one is struck by the broad range of
insecticides which are capable of doing this. In vitro incubation of
neurosecretory terminals in a variety of insecticides produces
alterations in electrical activity and ultrastructure, and the
release of bioassayable hormone. Neurosecretory cells are neurons
and possess all of the electrical properties of conventional neurons.
They also receive synaptic input from a variety of sources within the
central nervous system. Thus, neurosecretory cells may be activated
by insecticides which interfere with synaptic transmission and by
insecticides which interefere with nervous transmission. The susceptibility of neurosecretory systems may lie in their exposure to the
haemolymph. Neurosecretory terminals, and in some cases the entire
cell (eg. peripheral cells), have membranes in direct contact with
the haemolymph. The direct exposure of a cell membrane to insecticide will obviously make that cell more susceptible to an insecticide
than a cell lying within the central nervous system. In addition,
direct exposure of a cell membrane may produce changes quite apart
from the proposed mode of action of an insecticide. Thus, since
large areas of the membrane of neurosecretory cells are exposed
within neurohaemal organs, there may be changes in the normal activities of these areas.

Whether an insecticide has its direct action upon the central nervous system or upon the neurosecretory cells, a consequence appears to be release of hormone. Since neurohormones play a role in an enormous variety of metabolic, physiological and behavioural events, the liberation of a plethora of these controlling factors may be expected to upset the physiological balance of the insect and may contribute quite significantly to eventual death.

ACKNOWLEDGEMENTS

I am most grateful to Dr. Angela Lange for helpful suggestions throughout the writing of this review, and for the preparation of the manuscript.

REFERENCES

Brady, J., and Maddrell, S.H.P., 1967, Neurohaemal organs in the medial nervous system of insects, Z. Zellforsch. Mikrosk. Anat., 76:389-404.

Carlsen, J., Herman, W.S., Christensen, M., and Josefsson, L., 1979, Characterization of a second peptide with adipokinetic and red-pigment-concentrating activity from the locust corpora cardiaca, Insect Biochem., 9:497-501.

Casida, J.E., and Maddrell, S.H.P., 1971, Diuretic hormone release on poisoning Rhodnius with insecticide chemicals, Pestic. Biochem. Physiol., 1:71-83.

Colhoun, E.H., 1959, Some physiological and pharmacological effects of chlorinated hydrocarbons and organophosphorus poisoning, Proc. North Cent. Branch Entomol. Soc. Amer., 14:35-37.

Fifield, S., and Finlayson, L.H., 1978, Peripheral neurons and peripheral neurosecretion in the stick insect, Carausius morosus, Proc. R. Soc. Lond. B., 200:63-85.

Finlayson, L.H., and Osborne, M.P. 1968, Peripheral neurosecretory cells in the stick insect (Carausius morosus) and the blowfly larva (Phormia terraenovae), J. Insect Physiol., 14:1793-1801.

Finlayson, L.H., and Osborne, M.P., 1970, Electrical activity of neurohaemal tissue in the stick insect, Carausius morosus., J. Insect Physiol., 16:791-800.

Gabe, M., 1966, "Neurosecretion," Int. Ser. Mong. Biol., 28, Pergamon, New York.

Gainer, H., 1978, Input-ouput relations of neurosecretory cells, in: "Comparative Endocrinology," P.J. Gaillard and H.H. Boer, eds., Elsevier/North Holland Biomedical Press, Amsterdam.

Gersch, M., Richter, K., Bohm, G.A. and Sturzbecher, J., 1970, Selektive Ausschuttung von Neurohormonen nach elektrischer Reizung der Corpora Cardiaca von Periplaneta americana in vitro., J. insect Physiol., 16:1991-2013.

Gosbee, J.L., Milligan, J.V. and Smallman, B.N., 1968, Neural

properties of the protocerebral neurosecretory cells of the
adult cockroach, <u>Periplaneta</u> <u>americana</u>, <u>J. Insect Physiol.</u>,
14: 1785-1792.

Kater, S.B., 1968, Cardioaccelerator release in <u>Periplaneta</u> <u>americana</u>
(L.), <u>Science</u>, New York, 160:765-767.

Krogh, I.M. and Normann, T.C., 1977, The corpus cardiacum neurosecretory cells of <u>Schistocerca</u> <u>gregaria</u>. Electron microscopy
of resting and secreting cells, <u>Acta. Zool. (Stockh)</u>, 58:69-79.

Maddrell, S.H.P., and Gee, J.D., 1974. Potassium-induced release of
the diuretic hormones of <u>Rhodnius</u> <u>prolixus</u> and <u>Glossina</u> austeni:
Ca dependence, time course and localization of neurohaemal
areas, <u>J. Exp. Biol.</u>, 61:155-171.

Maddrell, S.H.P., and Nordmann, J.J., 1979, "Neurosecretion," Wiley,
New York.

Maddrell, S.H.P., and Reynold, S.E., 1972, Release of hormones in
insects after poisoning with insecticides, <u>Nature</u>, 236:404-406.

Moreteu, P.B., and Ramade, F., 1979, Lesions ultrastructurales des
corpora cardiaca d'un insects, <u>Locusta</u> <u>migratoria</u> (Orthoptere),
intoxique par de lindane, <u>Zool. Jb. Physiol.</u>, 83:340.

Normann, T.C., 1965, Neurosecretory system of an adult <u>Calliphora</u>
<u>erythrocephala</u>, 1. The fine structure of the corpus cardiacum,
with some observations on adjacent organs. <u>Z. Zellforsch</u>
<u>Mikrosk. Anat.</u>, 67:461-501.

Normann, T.C., 1980, Release of neurohormones in blowflies <u>Calliphora</u>
<u>vicinia</u> with respect to insecticidal action, <u>in</u>: "Insect
neurobiology and pesticide action," Soc. Chem. Ind., London.

Normann, T.C., and Samaranayaka-Ramasamy, M., 1977, Secretory
hyperactivity and mitochondrial changes in neurosecretory cells
of an insect, cellular effects of the insecticide lindane.
<u>Cell. Tiss. Res.</u>, 183:61-69.

Orchard, I., 1983, Neurosecretion: Morphology and Physiology, <u>in</u>:
"Endocrinology of Insects," R.G.H. Downer and H. Laufer, eds.,
Alan R. Liss, Inc., New York.

Orchard, I. 1982, Octopamine in insects. Neurotransmitter, neurohormone and neuromodulator. <u>Can. J. Zool.</u>, 60:659-669.

Orchard, I., 1980a, Electrical activity of neurosecretory cells and
its modulation by insecticides, <u>in</u>: "Insect Neurobiology and
Pesticide Action", Soc. Chem. Inc., London.

Orchard, I., 1980b, The effects of pyrethroids on the electrical
activity of neurosecretory cells from the brain of <u>Rhodnius</u>
<u>prolixus</u>, <u>Pestic. Biochem. Physiol.</u>, 13: 220-226.

Orchard, I., and Finlayson, L.H., 1976, The electrical activity of
mechanoreceptive and neurosecretory neurons in the stick insect
<u>Carausius</u> <u>morosus</u>, <u>J. Comp. Physiol.</u>, 107: 327-338.

Orchard, I., and Osborne, M.P., 1979, The action of insecticides on
neurosecretory neurons in the stick insect, <u>Carausius</u> <u>morosus</u>,
<u>Pestic. Biochem. Physiol.</u>, 10: 197-202.

Orchard, I., and Steel, C.G.H., 1980, Electrical activity of neurosecretory axons from the brain of <u>Rhodnius</u> <u>prolixus</u>: Relation
of changes in the pattern of activity to endocrine events during

the moulting cycle. brain Res., 191:53-65.

Orchard, I., and Loughton, B.G., 1985, Neurosecretion, in: "Comprehensive Insect Physiology, Biochemistry and Pharmacology," G.A. Kerkut and L.I. Gilbert, eds., Pergamon Press, Oxford (in press).

Orchard, I. and Loughton, B.G., 1981a, The neural control of release of hyperlipaemic hormone from the corpus cardiacum of Locusta migratoria, Comp. Biochem. Physiol., 68A:25-30.

Orchard, I., and Loughton, B.G., 1981b, Is octopamine a transmitter mediating hormone release in insects?, J. Neurobiol., 12:143-153.

Orchard, I., Friedel, T., and Loughton, B.G., 1981, Release of neurosecretory protein from the corpora cardiaca of Locusta migratoria induced by high potassium saline and compound action potentials, J. Insect Physiol., 27:297-304.

Orchard, I., Loughton, B.G., Gole, J.W.D. and Downer, R.G.H., 1983a, Synaptic transmission elevates adenosine 3', 5'-monophosphate (cyclic AMP) in locust neurosecretory cells. Brain Res., 258:152-155.

Orchard, I., Gole, J.W.D., and Downer, R.G.H., 1983b, Pharmacology of aminergic receptors mediating an elevation in cyclic AMP and release of hormone from locust neurosecretory cells, Brain Res., 288:349-353.

Orr, G.L., and Downer, R.G.H., 1983, Effects of reserpine and octopamine on lindane-induced changes in tissue carbohydrate levels in the American cockroach, Periplaneta americana L. Pestic. Biochem. Physiol., 19:151-156.

Osborne, M.P., 1979, The effect of gamma-HCH upon the ultrastructure of the neurohaemal organs on the median nerves of the stick insect Carausius morosus, Pestic. Sci., 10:320.

Samaranayaka, M., 1974, Insecticide-induced release of hyperglycemic and adipokinetic hormones of Schistocerca gregaria, Gen. Comp. Endocrinol., 24:424-436.

Samaranayaka, M., 1976, Possible role of involvement of monoamines in the release of adipokinetic hormone in the locust, Schistocerca gregaria., J. Exp. Biol., 65:415-425.

Samaranayaka, M., 1977, Role of acetylcholine in organophosphate-induced release of adipokinetic hormone in the locust, Schistocerca gregaria, Pestic. Biochem. Physiol., 7:283-288.

Samaranayaka, M., 1978, Insecticide-induced release of neurosecretory hormones, in: "Pesticide and Venom Neurotoxicity," D.L. Shankland, R.M. Hollingworth, and T.Smyth, Jr. eds. Plenum, New York.

Scharrer, B., and Kater, S.B., 1969, Neurosecretion XV. An electron microscopic study of the corpora cardiaca of Periplaneta americana after experimentally induced hormone release. Z. Zellforsch. Mikrosk. Anat., 95:177-186.

Singh, G.J.P. and Orchard, I., 1982, Is insecticide-induced release of insect neurohormone a secondary effect of hyperactivity of the central nervous system? Pestic. Biochem. Physiol., 17:232-

242.

Singh, G.J.P., and Orchard, I., 1983, Action of bioresmethrin on the corpus ardiacum of <u>Locusta migratoria</u>, <u>Pestic. Sci.</u>, 14:229–234.

Singh, G.J.P., Barker, J.F. and Kundu, S.C. 1982a, Bioresmethrin-induced alterations in the ultrastructure of neurosecretory cells of insect corpora cardiaca. <u>Pestic. Biochem. Physiol.</u>, 18:158–168.

Singh, G.J.P., Orchard, I., and Loughton, B.G., 1982b, Pharmacology of insecticide-induced release of hyperlipaemic hormone in the locust, <u>Locusta migratoria</u>, <u>Gen. Pharmac.</u>, 13: 471–475.

Sternberg, J., Chang, C.S., and Kearns, C.W., 1957, DDT-induced toxins in insect blood, <u>Fed. Proc.</u>, 16:124.

Stone, J.V., and Mordue, W., 1979, Isolation of granules containing adipokinetic hormone from locust corpora cardiaca by differential centrifugation, <u>Gen. Comp. Endocrinol.</u>, 39: 543–554.

CYCLIC AMP SYNTHESIS AND DEGRADATION: POSSIBLE TARGETS FOR PESTICIDE ACTION

James A. Nathanson

Department of Neurology, Harvard Medical School
Neuropharmacology Research Laboratory
Massachusetts General Hospital, Boston, MA 02114

INTRODUCTION

Cyclic AMP (Fig. 1), produced through activation of the membrane-bound enzyme, adenylate cyclase, mediates the intracellular actions of a number of vertebrate and invertebrate hormones and neurotransmitters. Disruption of cyclic AMP formation and/or degradation, therefore, has profound consequences for the orderly regulation of cellular physiological processes. Recent evidence, described below, suggests that through pharmacological intervention it is possible to alter tissue cyclic AMP levels in a variety of insects. Such alterations cause profound behavioral and metabolic abnormalities and, in some cases, disrupt feeding and cause insect death. Because certain hormone receptors involved in regulating cyclic AMP synthesis may exist primarily, if not exclusively, in invertebrates, it may be possible to develop new classes of "cyclic AMP-active" pesticides which have decreased toxicity for vertebrates. By coupling such pesticides with an inhibitor of cyclic AMP degradation, it should be possible to increase potency while maintaining invertebrate selectivity.

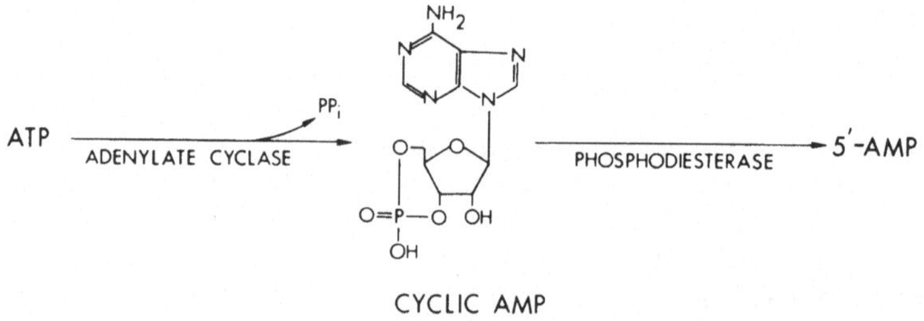

Fig. 1. Structure and enzymatic metabolism of cyclic AMP.

157

INTERACTIONS WITH ADENYLATE CYCLASE

Cyclic AMP Synthesis

In cells containing a particular hormone-coupled adenylate cyclase, receptor binding of the hormone causes a series of conformational changes in cell membrane proteins. These changes ultimately result in the activation of adenylate cyclase and the enzymatic conversion of ATP to cyclic AMP (Fig. 2) (see Nathanson and Kebabian, 1982, for review). As a general rule, cyclic AMP, formed by hormone-sensitive adenylate cyclase, exerts its effects on intracellular physiology by activating cyclic AMP-dependent protein kinases. These important regulatory enzymes stimulate the phosphorylation of intracellular proteins (Fig. 3) (Greengard, 1978; Nestler and Greengard, 1984).

Perhaps the best characterized hormone-sensitive adenylate cyclase found primarily in invertebrates (and not vertebrates) is the octopamine-stimulated adenylate cyclase originally described by Nathanson and Greengard (1973, 1974) (Nathanson, 1976; 1979a,b;

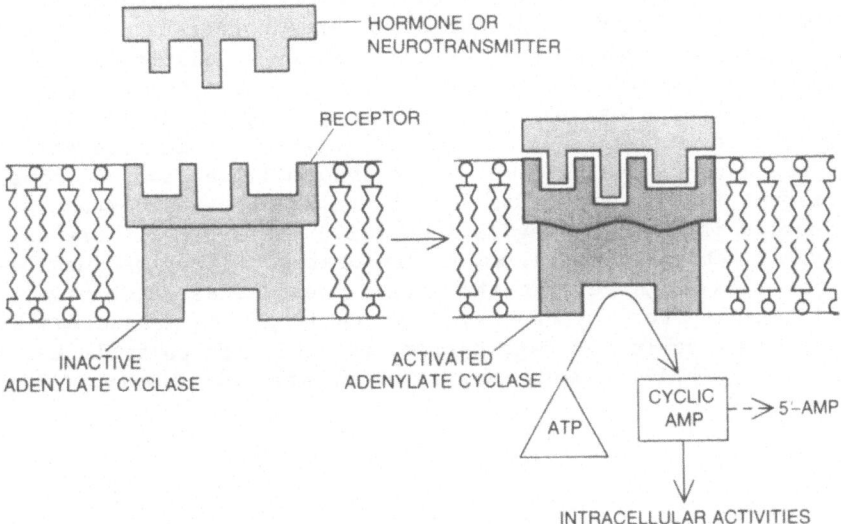

Fig. 2. Simplified schematic diagram showing the interaction of membrane receptors with extracellular hormone and membrane associated adenylate cyclase. Binding of hormone causes conformational changes in adenylate cyclase regulatory proteins (not shown) ultimately activating the catalytic subunit of adenylate cyclase and causing increased synthesis of intracellular cyclic AMP (from Nathanson and Greengard, 1977; copyright 1977 by Scientific American; all rights reserved).

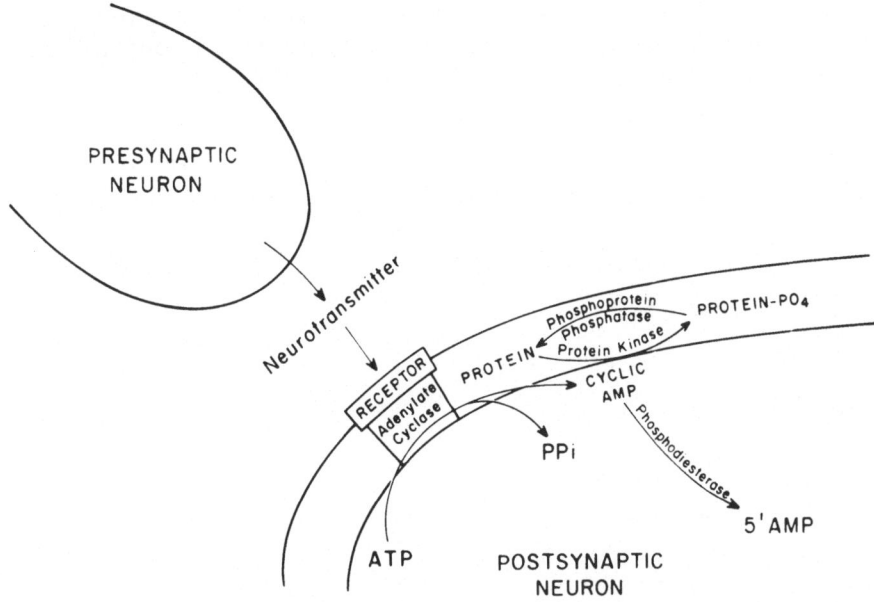

Fig. 3.   Following formation, cyclic AMP exerts its effects by
activating protein kinases. Some of these kinases (depicted
here) reside in postsynaptic membranes. Their activation
results in phosphorylation of membrane proteins, some of
which may be involved in regulating membrane permeability
(from Greengard, 1977, with permission).

Evans, 1980, Nathanson and Hunnicutt, 1981; Lingle et al., 1982;
Nathanson, 1985). This enzyme appears to mediate most, if not all,
of the physiological actions of octopamine which occur through type-2
octopamine receptors. Octopamine-sensitive adenylate cyclase can be
stimulated by octopamine as well as by certain synthetic agonists
which mimic the action of octopamine. As described below, such
synthetic agonists can disrupt insect behavior if administered in
large, non-physiological (pesticidal) doses.

Phenylethanolamines

4-Fluoro-phenylethanolamine (FEA) (an octopamine analog in
which fluorine has been substituted for the p-hydroxy group) has
greater lipid solubility than octopamine and is more resistant to
ring oxidation. FEA can stimulate octopamine-sensitive adenylate
cyclase with a maximum activity (V_{max}) close to that generated by
octopamine, although with lesser potency (Table 1) (Nathanson and
Hunnicutt, 1979b). When coupled with a phosphodiesterase inhibitor
(described below) and applied as a spray to natural food, FEA can
cause hyperactivity, tremors, and antifeeding effects in <u>Manduca
sexta</u> larvae (Nathanson, 1985).

Table 1. Structure-activity relationships of various octopaminergic
          agents and insecticide classes as agonists and antagonists
          of octopamine-sensitive adenylate cyclase

| Compound | Agonist activity | | Antagonist Activity |
	V_{max} (% OCT)	K_a Ratio	K_i (micromolar)
PHENYLETHANOLAMINES (PEA)			
Octopamine	100	1.	None
4-Fluoro-PEA	72	0.11	NT
FORMAMIDINES			
Chlordimeform (CDM)	9	0.4	300*
Monodemethyl-CDM	70	6.	300*
PEHNYLIMINOIMIDAZOLIDINES			
NC5	100	20.	None
NC7	70	9.	NT
CHLORINATED HYDROCARBONS			
DDT	NA	-	25
Dieldrin	NA	-	30
ORGANOPHOSPHATES			
Chlorpyriphos (Dursban)	NA	-	75*
Imidan (Phosmet)	NA	-	>180*
CARBAMATES			
Carbaryl (Sevin)	NA	-	360
Bendiocarb (Dexachlor)	NA	-	100
PYRETHROIDS			
Pyrethrum I, II	NA	-	225
Resmethrum	NA	-	>750*
OTHERS			
Dinocap (Karathane)	NA	-	25*
Rotenone	NA	-	75

V_{max} is the maximum activation of light organ adenylate cyclase by
an agonist (over the range 10^{-7} to 3×10^{-5}M) compared to the
activation by octopamine (=100%). K_a ratio is calculated by
dividing the K_a for octopamine by the K_a for the agonist. Values
>1 indicate compounds more potent than octopamine. K_i is the
calculated inhibitory constant for inhibition by the compound of
adenylate cyclase activity in the presence of 10^{-5}M octopamine.
*Substantial inhibition of basal enzyme activity, suggesting non-
receptor related effect. NA = no activity. NT = not tested.

Formamidines

Hollingworth and Murdock (1980) reported that the known pesticides, chlordimeform (CDM) and demethylchlordimeform (DMCDM), could mimic octopamine in causing light emission in the firefly lantern, a tissue known to be rich in octopamine-sensitive adenylate cyclase (Nathanson, 1979). Nathanson and Hunnicutt (1981) subsequently showed that DMCDM (the metabolic breakdown product of CDM formed <u>in vivo</u>) is a potent partial agonist of firefly octopamine-sensitive adenylate cyclase. DMCDM is 5 times as potent as octopamine and has a V_{max} of about 65% of that due to octopamine (Fig. 4). CDM, however, is a much weaker octopamine agonist than DMCDM (Table 1). This finding supports previous speculation that the toxic effects of CDM are actually due to its metabolic breakdown to DMCDM. Various other pharmacological data strongly suggest that the physiological and toxic effects of DMCDM are due to its activation of octopamine-sensitive adenylate cyclase. Further support for the theory, which links the interaction of formamidines with octopamine receptors, has come from electrophysiological studies of insect muscle and neuromuscular junction (Evans and Gee, 1980).

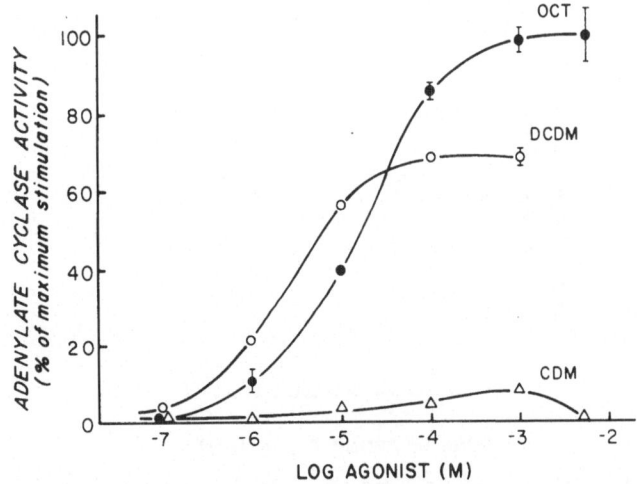

Fig. 4. Octopaminergic effects of demethylchlordimeform (DCDM). DCDM has a K_a for firefly light organ adenylate cyclase which is 3-4 fold lower than that for octopamine. However, as shown, DCDM is a partial agonist. It activates enzyme activity only about 70% as well as octopamine. Chlordimeform (CDM) has very little activity. This indicates that, <u>in vivo</u>, the insecticidal activity of CDM is probably due to its demethylation to DCDM (from Nathanson and Hunnicutt, 1981, with permission).

Phenyliminoimidazolidines

Recent studies indicate that certain phenyliminoimidazolidine derivatives are extremely potent agonists of octopamine-sensitive adenylate cyclase (Nathanson, 1984, 1985). The NC5 derivative, for example, is both fully active and 20-times more potent than octopamine in stimulating the firefly enzyme (Table 1). When applied as a spray to tomato leaves, NC5 coupled with a phosphodiesterase inhibitor greatly increases antifeeding activity. Recent studies show that the sensitivity of octopamine-2 receptors to activation by NC5 and various other synthetic octopamine agonists (above) varies among different insect species (Nathanson, 1985). Those species with the greatest enzyme sensitivity also show the greatest behavioral disruption after exposure to the agonists (Fig. 5).

Other Insecticide Groups as Activators of Adenylate Cyclase

In other experiments, described here, we have investigated the possible effects of several classes of known insecticides on the activity of octopamine-sensitive adenylate cyclase. A number of these compounds are known to affect cellular biochemistry in other ways (e.g., acetylcholinesterase inhibition, interaction with GABA

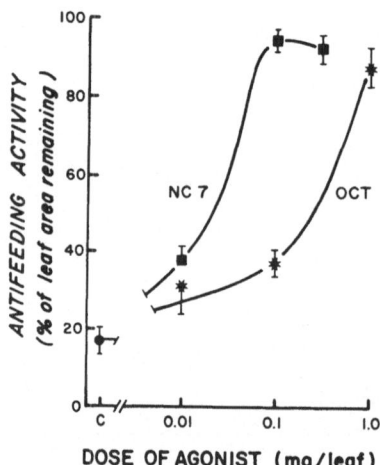

Fig. 5. The phenyliminoimidazolidines are potent octopaminergic agonists. In vitro tests (not shown) indicate that some derivatives are full agonists. Some are as much as 20-times more potent than octopamine itself. In vivo feeding tests (shown here) demonstrate that both octopamine and the phenyliminoimidazolidine (NC7) can exert antifeeding effects in Manduca 1st instar larvae. Compounds (combined with 0.1% IBMX) are applied as a spray to tomato leaves. Control leaves were treated with IBMX alone (from Nathanson, 1985, with permission).

receptors) but have not been well studied for their possible effects
on cyclic AMP-linked octopamine receptors.

The various compounds were tested for their ability to activate
adenylate cyclase activity in broken cell preparations of the firefly
light organ. As previously described, this tissue contains a high
concentration of octopamine- but not other hormone-stimulated
adenylate cyclases. Adenylate cyclase activity is normally measured
in a water-based buffer system. Unfortunately, however, many of the
insecticide groups being studied were highly insoluble in water.
Therefore, assay conditions were modified to include a solubilizing
agent such as dimethylsulfoxide (DMSO) or polyethylene glycol (PEG).
Extensive preliminary testing was done in order to determine the
highest concentration (approximately 20%) of DMSO or PEG which would
allow solubilization of the insecticide without excessively
inhibiting adenylate cyclase activity. Compounds were run in
parallel with octopamine to ensure that receptor sensitivity was
maintained.

Figures 6-9 (bottom curves) show that, at concentrations from
10^{-7}M, none of the 10 pesticides tested were able to activate
octopamine-sensitive adenylate cyclase. Three compounds (karathane,
chlorpyriphos [Dursban], and imidan) caused substantial inhibition of
basal enzyme activity when present at concentrations of 10^{-4} to 10^{-3}M. Other studies show that chlordimeform also causes a non-specific
inhibition of adenylate cyclase (Nathanson and Hunnicutt, 1981).
Because hormone-sensitive and basal adenylate cyclase activity was
inhibited (see below), it is likely that the inhibition observed was
not related to receptor-adenylate cyclase interaction. Furthermore,

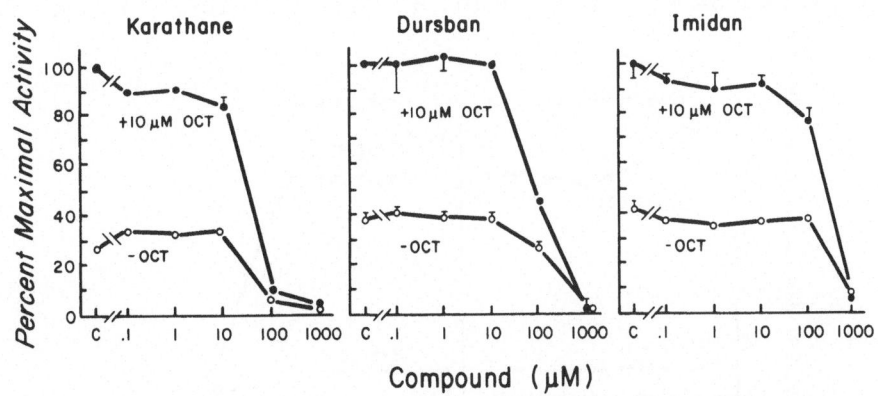

Fig. 6.   Effect of various insecticides on firefly light organ
          adenylate cyclase activity. Activity was measured in the
          presence of insecticide alone (bottom lines). The latter
          evaluated the ability of the insecticides to act as
          antagonists of octopamine-sensitive adenylate cyclase.

the high concentrations required to demonstrate inhibition suggest
that inactivation of adenylate cyclase activity by these pesticides
is unlikely to play a role in their insecticidal activity. Finally,
since the firefly light organ contains, primarily, an octopamine-
sensitive adenylate cyclase, these results do not preclude the
possible interaction of the above compounds with other hormone-
sensitive adenylate cyclases that are known to be present in other
insect tissues (Lingle et al., 1982).

Insecticides as Antagonists of Octopamine-sensitive Adenylate Cyclase

The various insecticide groups described above were also
evaluated for their ability to act as antagonists of the octopamine-2
receptor. Receptor blockade was measured as the ability to inhibit
activation of light organ adenylate cyclase by 10^{-5}M octopamine, a
dose approximately equal to the K_a of octopamine for its receptor.

Figures 6 and 7 (top curves) show that karathane, dursban,
imidan, and pyrethrum, at concentrations from 10^{-4} to 10^{-10}M,
inhibited octopamine-stimulated activity. However, as described
above, these compounds also proportionately inhibited basal enzyme
activity (bottom curves). Therefore, the inhibition observed was not
due to interaction with the octopamine receptor.

Resmethrum (Fig. 7) and the carbamate, carbaryl (Sevin, Fig. 8)
had relatively little effect on octopamine-stimulated activity.
However, bendiocarb, another carbamate, was able to selectively
inhibit octopamine activation, with an ED_{50} of about 10^{-10}M.
Rotenone and the chlorinated hydrocarbons, DDT and dieldrin (Fig. 9),
were the most potent antagonists among those insecticides tested,
with EC_{50} values between 10^{-5} and 10^{-4}M. The inhibitory constants for

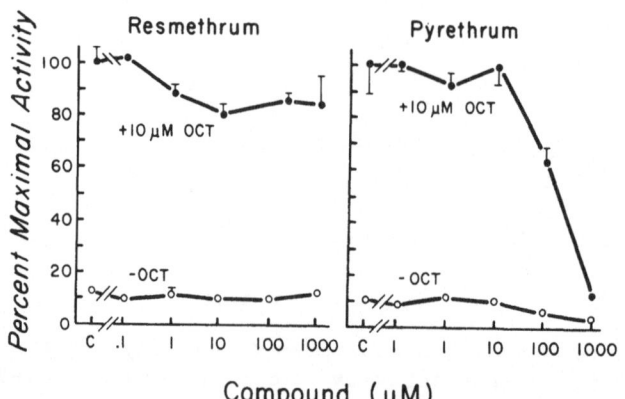

Fig. 7. Effect of pyrethrum-type insecticides on firefly adenylate
cyclase activity. See Fig. 6 for details.

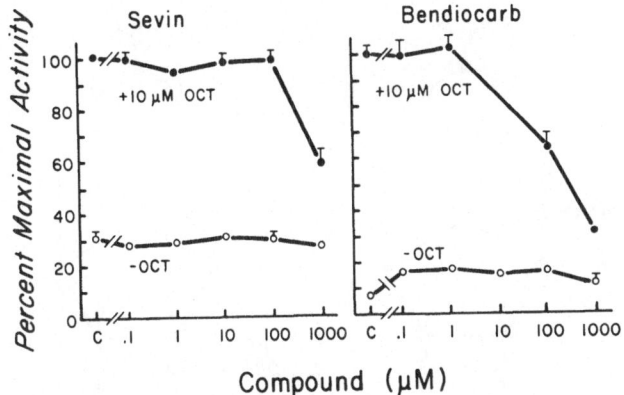

Fig. 8. Effect of carbamate-type insecticides on firefly adenylate
         cyclase activity. See Fig. 6 for details.

all the other compounds are summarized in Table 1.

INTERACTIONS WITH PHOSPHODIESTERASE

Termination of Cyclic AMP Action

The physiological processes initiated by cyclic AMP and cyclic
AMP-stimulated protein phosphorylation are terminated in two ways:
first, through the dephosphorylation of proteins by the enzyme, phosphoprotein phosphatase; second, through a decrease in intracellular
cyclic AMP levels. This latter mechanism is mediated primarily by the
enzyme, cyclic nucleotide phosphodiesterase, which converts cyclic AMP
to the inactive compound, 5'-AMP (Fig. 10). A small amount of cyclic
AMP may also be removed from cells by active transport.

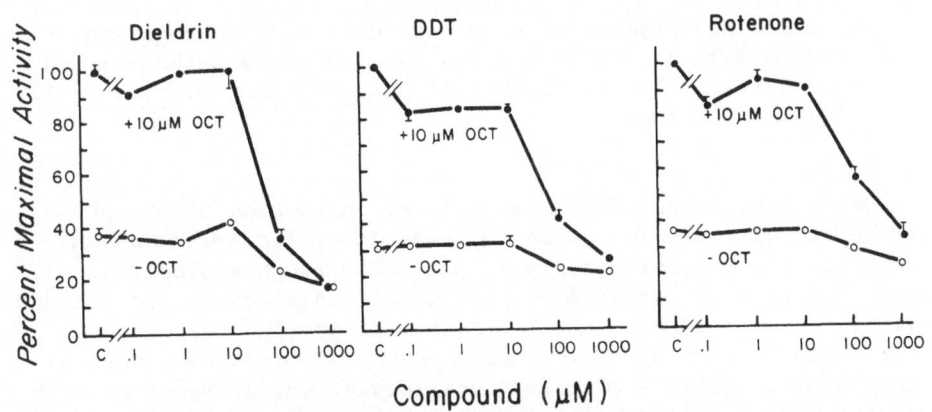

Fig. 9. Effect of additional insecticides on firefly adenylate
         cyclase activity. See Fig. 6 for details.

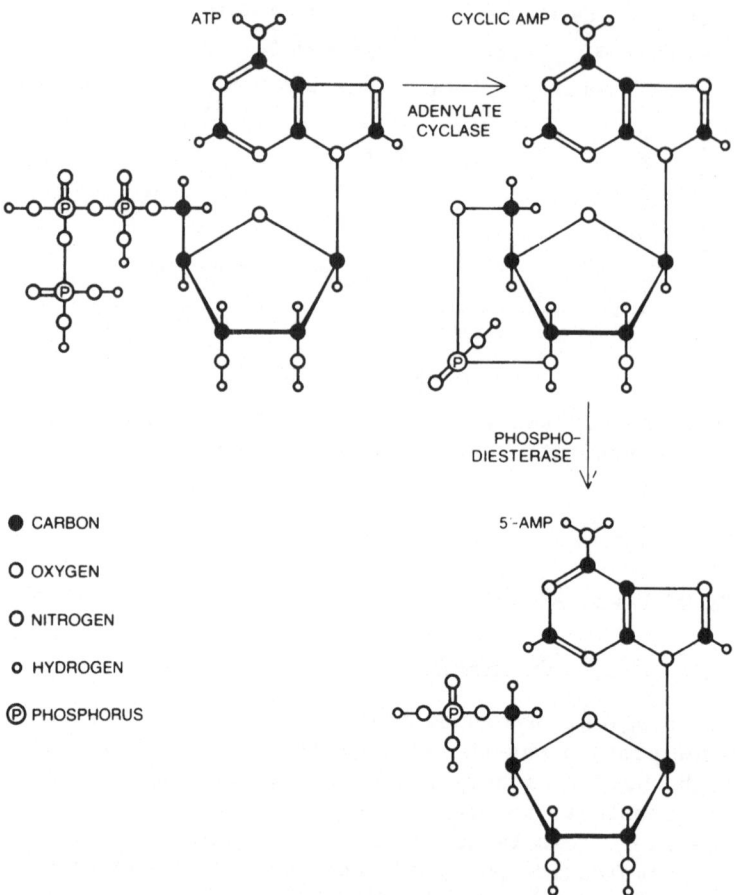

ATP

CYCLIC AMP

ADENYLATE
CYCLASE

PHOSPHO-
DIESTERASE

● CARBON

○ OXYGEN

○ NITROGEN

o HYDROGEN

℗ PHOSPHORUS

5'-AMP

Fig. 10. Once formed by adenylate cyclase, cyclic AMP is broken down
by cyclic nucleotide phosphodiesterase (PDE), which converts
cyclic AMP to 5'-AMP. Inhibition of PDE activity (such as
by caffeine) results in an enhancement and prolongation of
hormone effects mediated by cyclic AMP (from Nathanson and
Greengard, 1977; copyright 1977 by <u>Scientific</u> <u>American</u>, all
rights reserved).

In cells containing adenylate cyclase, inhibition of phosphodiesterase activity results in an increase in cyclic AMP content.
This increase concentration augments and prolongs physiological
processes mediated by cyclic AMP. Because of this action, we decided
to investigate what effects the inhibition of phosphodiesterase
activity would have on the pesticidal potency of compounds, such as
the octopamine-2 agonists described previously, which appear to work
through stimulation of adenylate cyclase. According to our
hypothesis, phosphodiesterase inhibitors should act as synergists and
increase the pesticidal activity of the primary compounds (Nathanson,

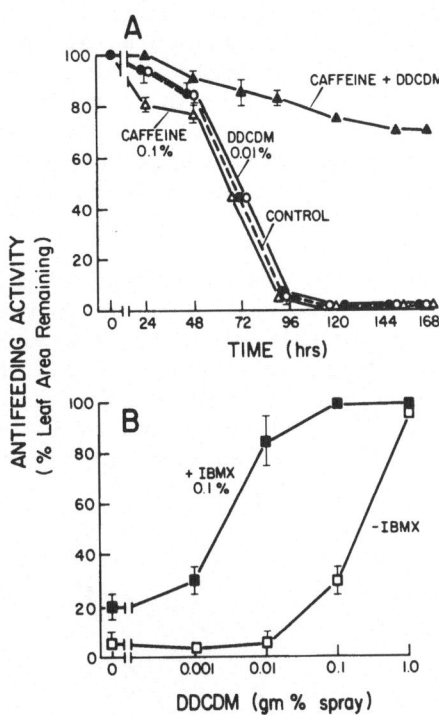

Fig. 11. Synergistic effects of methylxanthine PDE inhibitors on antifeeding effects of formamidines. (A) Time course experiment
showing synergistic effect of caffeine and DDCDM, compared
with each compound alone, on consumption of tomato leaves
sprayed with various agents. (B) Dose-response study showing
large increase in potency of DDCDM when combined with fixed
0.1% concentration of the PDE inhibitor, IBMX (from
Nathanson, 1984, with permission).

1984). Furthermore, if such enhancement did occur it would serve as
further evidence that the pesticidal activity of octopamine agonists,
themselves, is mediated through cyclic AMP production.

      Initial studies were carried out with phosphodiesterase inhibi-
tors of the methylxanthine class (Nathanson, 1984). Fig. 11A shows
the effect of combining a low (subtoxic) dose of the octopamine
agonist, 1,3,-7-trimethylxanthine (caffeine). In these studies,
compounds were applied as a spray to tomato leaves. Then, first
instar Manduca sexta larvae were placed on the leaves and allowed to
feed for five days. Fig. 11A, which plots the leaf area remaining
after each day, shows that the low doses of caffeine or DDCDM alone
had no effect on feeding compared with control leaves. However, when
combined, the two compounds produced a marked antifeeding effect. In
other experiments (Fig. 11B), we investigated the effects of a fixed

dose of caffeine on the dose-dependent toxicity of a series of DDCDM
concentrations. The resulting data showed a leftward shift in the
antifeeding dose-response curve of DDCDM indicating an increase in
potency. In various experiments, this increase in potency varied
between 5 and 50-fold. Similar increases in potency were seen when
caffeine was added to other octopamine agonists, such as CDM, 4-
fluoro-phenylethanolamine, NC5, and octopamine, itself (Nathanson,
1984).

    In other studies, we found that a variety of phosphodiesterase
inhibitors were effective in increasing the potency of the octopamine
agonist. Furthermore, by directly measuring the inhibition of phosphodiesterase activity in broken cell preparations of Manduca nerve
cords, we were able to demonstrate that there is an excellent
correlation between the ability of compounds to inhibit phosphodiesterase and their ability to act as antifeeding synergists.

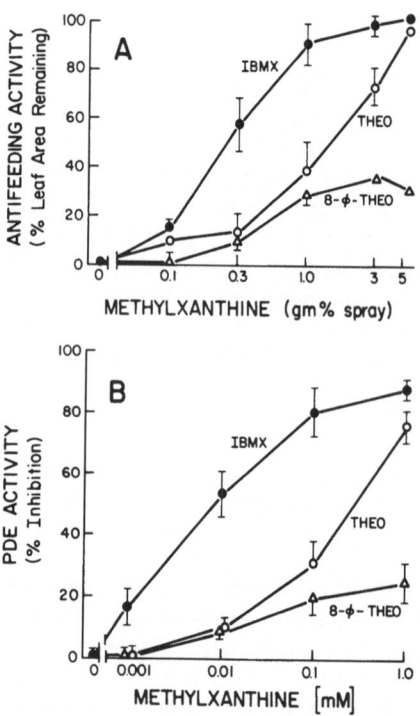

Fig. 12. (A) Antifeeding effects of various methylxanthines applied as
          a spray to tomato leaves fed to 1st instar tobacco hornworm
          larvae. (B) Effect of same compounds on inhibiting PDE
          activity in broken cell preparations of hornworm nerve cord.
          Note the correlation between the two graphs (in part, from
          Nathanson, 1984, with permission).

At higher doses, the methylxanthines, themselves, could inhibit feeding even in the absence of an adenylate cyclase agonist. Furthermore, the ability of various compounds to disrupt feeding correlated well with their ability to inhibit PDE activity in Manduca nerve cord (see Fig. 12, for example). In vitro studies revealed that the antifeeding effect of the methylxanthines (alone or in combination with an octopamine agonist) was associated with an increase in cyclic AMP content in isolated Manduca nerve cord (Table 2).

DISCUSSION

These results provide strong support for the hypothesis that the pesticidal effects of octopamine-2 agonists are mediated through stimulation of cyclic AMP formation in insect tissues. Furthermore, they raise the possibility that phosphodiesterase inhibitors may be useful as synergists for the formamidines, the phenyliminoimidazolidines, and, potentially, for any pesticide working through activation of a receptor-associated adenylate cyclase. Since the methylxanthines alone also caused antifeeding effects, potent phosphodiesterase inhibitors, by themselves, should be considered for pest control, too. Indeed, it is of interest that the concentration of the phosphodiesterase inhibitor, caffeine, present naturally in fresh tea leaves, is sufficient to kill Manduca larvae. This observation raises the possibility that caffeine and other methylxanthines, which are found in rather high concentrations in several plant species, may function as endogenous pesticides (Nathanson, 1984).

Table 2.   Effect of various octopamine agonists, in the absence or presence of the PDE inhibitor IBMX, on cyclic AMP content of Manduca nerve cord

Compound	Cyclic AMP (% control)	
	−IBMX	+IBMX (0.1mM)
Didemethylchlordimeform (0.1mM)	540 ± 380	1110 ± 270
NC7 (0.1mM)	220 ± 20	1390 ± 300
Octopamine (0.1mM)	90 ± 30	640 ± 100
IBMX alone	−	220 ± 30

Isolated Manduca nerve cord segments were incubated in oxygenated insect saline for 10 minutes. After boiling for 2 minutes, segments were homogenized, centrifuged, and the supernatants assayed for cyclic AMP content by protein binding assay. The values shown (mean ± SEM for N=3) are expressed as a percent of control activity (6.25 ± 1.05 pmol/mg protein/min) (from Nathanson, 1984).

ACKNOWLEDGEMENTS

    I thank E.J. Hunnicutt and C.J. Owen for technical assistance.
This work was supported, in part, by the JLN-Daniels Research Fund and
the McKnight Foundation.

REFERENCES

Evans, P.D., 1980, Biogenic amines in the insect nervous system,
    Adv. Insect Physiol., 15:317-473.
Evans, P.D. and Gee, J.D., 1980, Action of formamidine pesticides on
    octopamine receptors, Nature, 270:60-62.
Greengard, P., 1978, "Cyclic Nucleotides, Phosphorylated Proteins, and
    Neuronal Function," Raven Press, New York.
Hollingworth, R.M. and Murdock, L.L., 1980, Formamidine pesticides:
    octopamine-like actions in a firefly, Science, 208:74-76.
Lingle, C.J., Marder, E., and Nathanson, J.A., 1982, The role of
    cyclic nucleotides in invertebrates, in: "Cyclic Nucleotides
    II," J. Kebabian, and J. Nathanson, eds., Springer Verlag, New
    York.
Nathanson, J.A., 1976, Octopamine-sensitive adenylate cyclase and its
    possible relationship to the octopamine receptor, in: "Trace
    Amines and the Brain," E. Usdin and M. Sandler, eds. Marcel
    Dekker, New York.
Nathanson, J.A., 1979, Octopamine receptors, adenosine 3',5'-
    monophosphate, and neural control of firefly flashing, Science,
    203:65-68.
Nathanson, J.A., 1984, Caffeine and related methylxanthines: possible
    naturally occurring pesticides. Science, 226: 184-187.
Nathanson, J.A., 1985, Characterization of octopamine-sensitive
    adenylate cyclase: development of a class of potent and selec-
    tive octopamine-2 receptor agonists with toxic effects in
    insects, Proc. Natl. Acad. Sci, (USA), in press.
Nathanson, J.A., and Greengard, P., 1973, Octopamine-sensitive
    adenylate cyclase: evidence for a biological role of octopamine
    in nervous tissue, Science, 180:308-310.
Nathanson, J.A., and Greengard, P., 1974, Serotonin-sensitive
    adenylate cyclase in neural tissue and its similarity to the
    serotonin receptor: a posible site of action of lysergic acid
    diethylamide, Proc. Natl. Acad. Sci., 71: 797-801.
Nathanson, J.A., and Greengard, P., 1977, Second messengers in the
    brain, Scientific American, 237: 108-119.
Nathanson, J.A., and Hunnicutt, E.J., 1979a, Neural control of light
    emission in Photuris larvae: identification of octopamine-
    sensitive adenylate cyclase, J. Exp. Zool., 208:255-262.
Nathanson, J.A., and Hunnicutt, E.J., 1979b, Octopamine-sensitive
    adenylate cyclase: properties and pharmacological characteri-
    zation, Soc. Neurosci. Abstr., 5:346.

Nathanson, J.A., and Hunnicutt, E.J., 1981, N-demethylchlordimeform:
    a potent partial agonist of octopamine-sensitive adenylate
    cyclase, <u>Mol. Pharmacol.</u>, 20:68-75.
Nathanson, J.A., and Kebabian, J.W., 1982, Eds. "Cyclic Nucleotides
    I.," Springer Verlag, New York.
Nestler, E.J., and Greengard, P.G., 1984, "Protein Phosphorylation in
    the Nervous System," Wiley, New York.

ON INHIBITORY ACTION OF DDT AND PYRETHROIDS ON ATP-UTILIZING, CALCIUM TRANSPORTING SYSTEMS IN THE NEURAL TISSUES

Fumio Matsumura

Pesticide Research Center
Michigan State University
East Lansing, Michigan 48824-1311   USA

INTRODUCTION

As early as 1974 a suggestion was made that DDT's phenomenal insecticidal activities may be related to its ability to antagonize the effect of calcium (Welsh and Gordon, 1947). Also some scientists have noted at a very early stage that DDT affects mitochondrial functions, particularly those involved in ATP synthesis (Gonda et al., 1959; O'Brien, 1967). These and other incipient or sporadic observations have not attracted enough attention of toxicologists probably because of the primitive state of background biochemical knowledge and the lack of a logical explanation of the events.

Throughout the 1950's and 1960's the major research activities and the accomplishments in the field of action mechanisms of insecticides were made by electrophysiologists who found striking effects of DDT on the nervous system, particularly on the peripheral axons, to increase the frequency of evoked and spontaneous action potentials (see Kearns, 1956; Narahashi, 1976; Woolley, 1982). Subsequent studies have revealed that DDT mainly affects Na^+ channel by interfering with its inactivation mechanisms (Narahashi, 1976). Later studies with pyrethrin and its synthetic analog allethrin indicated that, while there are minor differences, the basic cause of their actions appears to be similar to that of DDT (Narahashi, 1979).

While the progress in electrophysiological science has been phenomenal and, with the development of voltage clamp techniques and now the patch clamp technique, the knowledge of such approaches provided in this field has been very valuable, the lack of biochemical understanding of the events involved in DDT's and now pyrethroids' action on the nervous system has hampered the progress in this field.

In the absence of a proper explanation some scientists have
resorted to postulations such as DDT's molecular fitting to inter-
lipid lattice, or physical interference by plugging into the NA^+
channel, etc. These speculations have not helped the matter as they
have not provided any experimental or theoretical evidence or basis
on which future studies could be built.

REVIEW OF RELATED WORK

Earlier Studies on DDT's Effects on ATPases

In 1969 it was discovered by two groups of scientists that DDT
at rather low concentrations inhibits neural ATPases (Matsumura and
Patil, 1969; Koch, 1969). The ATPases these workers studied were
mainly Na-K ATPase and Mg-ATPase. The former enzyme was becoming
known to function as a sodium pump at that time. Since it is
inhibited by a specific inhibitor, ouabain (a cardiac glycoside), its
entity was definable. The latter ATPase(s) activities, on the other
hand, represent sums of all ATP-hydrolyzing systems that are activated by Mg^{2+}. Since most of the ATP-utilizing systems require Mg^{2+},
it was not possible at that time to determine the identity of a
system or systems which are particularly sensitive to DDT and other
chlorinated hydrocarbon insecticides.

The possibility that Na-K ATPase is the target of DDT has been
eliminated at this stage (Matsumura, 1970), since inhibition of this
enzyme a priori is not expected to influence the membrane conductance
changes themselves. Therefore, the focus of attention of these
scientists has shifted to other neural ATPases which show high sensitivities to DDT.

The group led by Dr. Robert Koch and by Dr. Larry Cutkomp has
begun an indepth examination of mitochondrial ATPases as a possible
target of chlorinated hydrocarbon insecticides. Their works have
culminated to reveal the extreme sensitivity of the F-1 unit ATPase
toward these insecticidal chemicals (Cutkomp et al., 1982).

Our research team, on the other hand, has taken the approach to
identify the ATPase in the axonic plasma membrane (axolemma) which
appears to be very sensitive to DDT and insensitive to DDE. We knew
that the enzyme resembles Na-K ATPase, since it is stimulated by Na^+
and K^+ (Bratkowski and Matsumura, 1972; Matsumura et al., 1969). The
rationale of this approach was that if such an enzyme is related to
the action of DDT, it must have some sensitivity to Na^+ and
furthermore must be located at the axolemma. The initial efforts
were made to maximize the activity of this "Na^+-K^+ sensitive ATPase"
by varying various ionic and other incubation conditions (Bratkowski

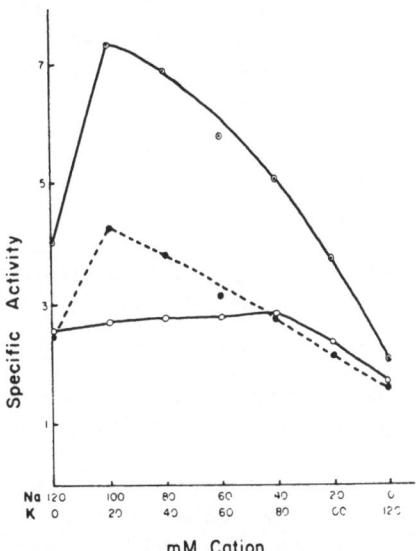

Fig. 1. Effects of ion combinations on the ATPase activity of the acetone powder preparation of the nerve ending. Specific activity is expressed as μmoles of Pi liberated/mg of protein per hr. Dotted circles with a solid line, control activity with no inhibitor; open circle with solid line, activity in the presence of 10^{-4}M ouabain; closed circles with dashed line, activity in the presence of 10^{-5}M DDT. (From Bratkowski and Matsumura, 1972).

and Matsumura, 1972). Also various inhibitors such as ouabain, cyanide, etc. were tested to eliminate unwanted ATPases in the system which have interfered with our tests (Fig. 1).

An important decision was made at this stage to keep the concentration of ATP low. The reason for this decision was to simulate the conditions of nerve preparations commonly employed by electrophysiologists to study action of chemicals on axons. For instance, it has been shown by other scientists that the concentration of ATP in the dialyzed squid giant axons (i.e., the axoplasm of the preparation is removed and replaced with an artificial K^+ rich internal saline solution. Such a preparation shows all normal electric properties) is in the order of 10^{-6}M (Mullins and Brinley, 1967). Repeated perfusion and washing do not seem to reduce the level of ATP below this level. At that time allethrin has been already shown to work on the squid axon, internally perfused in a similar manner to DDT (Narashahi, 1976). It has been reasoned that to study normal events of the excitable axonic membrane, one should not artificially elevate the ATP level beyond such a level. To achieve this objective, highly radioactive gamma ^{32}P-ATP preparations at levels of 10^{-7}M to 10^{-8}M were used. Doubtlessly fresh nerve membrane preparations contain much higher levels of endogenous ATP. Therefore, studies on the fate of the above labelled substrate should indicate the natural state of membrane constituents which interact with ATP. Subsequent studies (Doherty and Matsumura, 1974) have shown that the characteristics of the DDT-sensitive "ATPase" manifest well under such experimental conditions. More important was their finding that DDT increases the level of phosphorylated proteins at low ATP concentrations (i.e., 10^{-6} to 10^{-7}M). The results indicate that the system under study is likely to be an ATP-dependent phosphorylation/dephosphorylation system. Also interesting was the observation that calcium antagonized the process (Fig. 2).

In 1979 Ghiasuddin and Matsumura found that, when Ca^{2+} was used in place of Mg^{2+}, the bulk of the ATP-hydrolyzing activity becomes sensitive to DDT (Fig. 3). At the total protein level of 10 μg in total reaction volume of 1 ml the entire ATP-hydrolyzing activity becomes sensitive to DDT. The I_{50} value obtained in this manner was about 10^{-9}M (Fig. 4). The system's sensitivity toward DDE was much less. This discovery was a key event, leading to the realization of the importance of Ca^{2+} transporting systems in the nervous system in the action mechanisms of DDT and pyrethroids.

The above study did not prove that DDT-sensitive system does not require Mg^{2+}. It simply showed that it requires Ca^{2+}. Since there are a smaller number of Ca^{2+}-activated ATP-utilizing systems than Mg^{2+} activated ones, elimination of exogenously added Mg^{2+} from the assay medium greatly suppressed activities of unwanted ATPase which are activated solely by Mg^{2+}.

One of the ATP-dependent Ca^{2+} transporting systems which became known in the 1970's was Ca-Mg ATPase. The inhibitory action of DDT

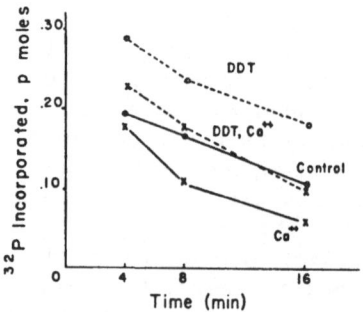

Fig. 2.   Time dependence of ^{32}P incorporation at 37°C. The buffer was of standard composition (60 mM Na^+, 60 mM K^+, 10 mM Mg^{2+}, pH 7.0) and the crude supernatant (1000xg) was used as the enzyme source. Ca^{2+} was 4 mM and DDT was 10^{-5}M. (From Doherty and Matsumura, 1974).

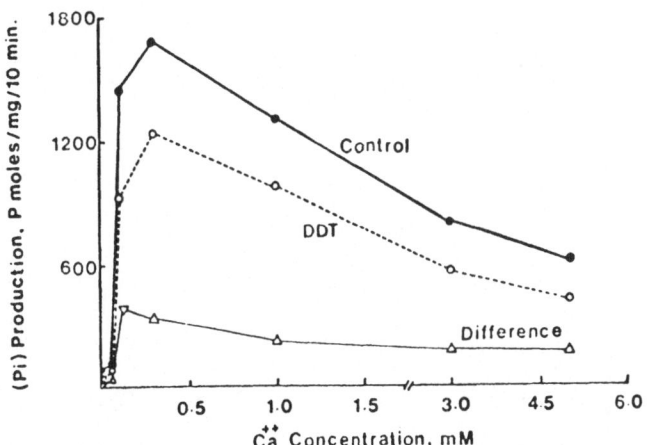

Fig. 3. Effect of Ca^{++} concentration on ATPase activity of lobster nerve membrane preparation. The standard assay conditions were 60 mM Na^+, 60 mM K^+, pH 7.3, 0.1 mM ouabain, 8 x 10^{-8} M ^{32}P ATP and temperature 30°C. Difference between "control" and "DDT" (10^{-5}M) treated is the indication of the level of the activity of DDT sensitive Ca-ATPase. (From Matsumura and Ghiasuddin, 1979).

on this ATPase was quickly established (Ghiasuddin and Matsumura, 1981). The Ca-Mg ATPase from the lobster axonic membrane may be inhibited to the extent of 95.6% with 10^{-5}M of DDT under the experimental condition. The I_{50} appeared to be in the order of 3 x 10^{-6}M. Two pertinent observations made in this study were (a) inhibition of Ca-Mg ATPase is more significant at higher temperatures (e.g., 30°C) and (b) its specific activity (i.e., percent ATP hydrolyzing activity due to Ca-Mg ATPase in the total ATPase activity) manifests better at high ATP concentration. These observations indicated that this enzyme was not the DDT sensitive "ATPase" (then designated as Ca-ATPase in the absence of a better term) we sought earlier, though Ca-Mg ATPase has been, no doubt, a factor present in our normal assay and is an important Ca-transporting ATPase.

Therefore, the nagging question on the identity of the DDT-sensitive Ca-ATPase persisted. One time it was thought to be similar to ecto-ATPase, but the idea was dropped quickly when Ghiasuddin and Matsumura (1981) found that in the presence of Mn^{2+}, which is an active replacement of Mg^{2+} or Ca^{2+} for ecto-ATPase, there was little effect of DDT on ATPase activity in the lobster muscle system.

Identity of DDT-sensitive Ca-ATPase

In 1980 to 1982 Clark and Matsumura (Clark, 1982; Matsumura and Clark, 1982) decided that probably the best way to identify the

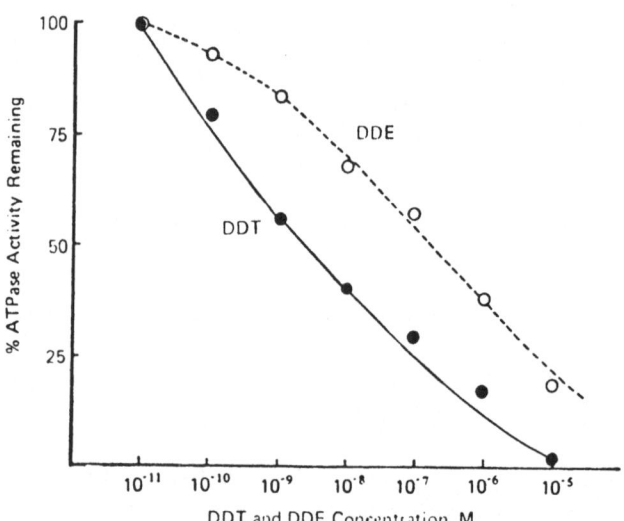

Fig. 4. Dose-response effect of DDT and DDE on Ca-ATPase activity of
the nerve membrane preparation. A low protein concentration
(4 μg/assay) was employed in this experiment under the standard assay condition. (From Matsumura and Ghiasuddin, 1979).

nature of the DDT-sensitive Ca-ATPase is to study its characteristics
in the squid axons of which physiological properties have been
exhaustively studied. For this purpose, the retinal nerve fibers
rich with very small axons were used. A new methodology had to be
developed for this project. As a result of extensive studies, we
have determined the identity of the Ca-ATPase is likely to be ATP
dependent Na/Ca exchange system (Matsumura and Clark, 1982). The
role of ATP is to increase the affinity of the exchange system to
intracellular Ca^{2+} (Fig. 5). The details of the experiments leading
to such a conclusion will be presented by Dr. Clark as a chapter in
this book, and, therefore, in this chapter only a few salient features of the discovery which relate to the understanding of DDT's and
pyrethroid's action will be mentioned.

Na/Ca exchange is the major mechanism by which cells extrude
intracelluarly accumulated Ca^{2+} into the surrounding medium. Its
major source of energy is the gradient of Na^+. Therefore, for each
mole of Ca^{2+} expelled, 3 moles of Na^+ must come into the cell,
according to the steep gradient of the latter ion across the axolemma. Despite such a reliance on Na^+ gradient for its energy,
enough evidence is accumulating to show that ATP participates in this
process, particularly at low intracellular Ca^{2+} concentrations
(DiPolo and Beauge, 1983; Blaustein, 1974). The action of ATP is
apparently to phosphorylate the exchange protein which makes it more
sensitive to $[Ca^{2+}]_i$ (i.e., an increase in affinity toward intra-

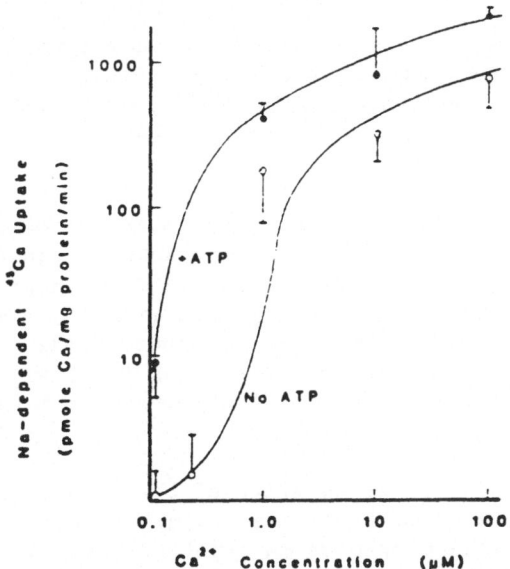

Fig. 5.   Effect of ATP on the relationship between Ca^{2+} concentration
and Na-dependent ^{45}Ca uptake (i.e., calcium load) by pre-
equilibrated retinal nerve microsomes of the squid, Loligo
pealei. A microsomal vesicle fraction was collected from
optic nerve axons by differential centrifugation utilizing
the method of Matsumura and Clark (1980) and resuspended into
a standard buffer solution (SBS: 50 mM $MgCl_2$, 1 mM ouabain, 2
mM KCN and 30 nM Tris base adjusted to pH 7.8) containing 425
mM NaCl and 10 mM KCl. After 2 hr pre-equilibration period
at $4^{\circ}C$, a 10 µl aliquot containing 31 µg protein (Lowry et
al., 1951) was added to 150 µl of a K-containing loading
solution (i.e., 425 mM KCl and 0.1 mM EGTA in SBS at $20^{\circ}C$).
Tris adenosine 5'-triphosphate (Sigma A0270) was added to
some loading solutions (+ ATP) at a concentration of 0.1 mM
and was absent from others (No ATP). Various calcium con
centrations were established at nominal values based on a Ca
EGTA stability constant value of $7.6 \times 10^6 M^{-1}$ (Portzehl et
al., 1964). Approximately 7 µCi of $^{45}CaCl_2$ (New England
Nuclear) was added to each buffered sample giving a specific
activity range for ^{45}Ca of 0.2-2.0 Ci/mmole Ca^{2+}. Calcium
influx continued for 5 min. at which time a 120 µl aliquot
was taken and placed into 8 ml of a non-effluxing, choline-
containing solution (i.e., 425 mM choline chloride, 10 mM
KCl, 0.5 mM EGTA in SBS). At the completion of a 2 min.
period, the entire suspension was vacuum filtered on pre-
washed 0.45µ millipore cellulose acetate filters. Each
filter was rinsed with 10 ml ice-cold choline-containing
solution. The washed filters were dissolved in liquid scintillation counting solution and the amount of ^{45}Ca remaining
in the microsomes was determined by liquid scintillation
spectrophotometric means. (From Matsumura and Clark, 1982).

cellular Ca^{2+}) and thereby more efficient in scavenging small rem-
nants of $[Ca^{2+}]_i$ in the cell. It must be mentioned here that the
maintenance of a very low intracellular Ca^{2+} level (10^{-8} to 10^{-9}M)
and thereby a steep Ca^{2+} gradient appears to be an essential requirement of excitable cells (DiPolo, 1973).

The above discovery brought one key realization on our part;
that is, the DDT-sensitive Ca-ATPase is not actually an ATPase, but
is a protein kinase-phosphatase system specialized in phosphorylating
and dephosphorylating the Na/Ca exchange protein. This was the very
reason it operates well with very small amounts of ATP (Doherty and
Matsumura, 1974) and why it is still active at low temperatures. A
term Na-Ca protein kinase-phosphatase (PKP) has been coined to describe the system.

There have been debates as to which of the calcium transporting
systems, Na/Ca exchange or Ca-Mg ATPase, is more important, and why
there are two ATP-utilizing systems in expelling calcium from the
cells. The answer might be that both systems are needed, and that,
depending upon tissues and circumstances, one system might become
more important than the other, since they have quite different characteristics. For instance, Na/Ca exchange may be more important at
lower temperatures and where $[Na^+]_o$ is more accessible, while Ca-Mg
ATPase (= Ca-pump) functions better at high temperatures and at
locations where a high level of ATP and only low levels of Na^+ are
available.

Studies on Mechanism of DDT-Pyrethroid Resistance (Kdr) in Insects

From a completely different point of view, we have begun a
research project to find a biochemical cause(s) for the acquired
nerve insensitivity (i.e., Kdr factor) by resistant insects to DDT
and pyrethroids. These are the insects which have developed resis-
tances to DDT, but show high levels of cross-resistance to pyrethroids, particularly to those belonging to type I group (Scott and
Matsumura, 1981, 1983; also see Georghiou and Saito, 1982). Apparently the cause for resistance is not related to metabolism or
transport of the insecticides. Since pyrethroids are very important
insecticides, there has been a serious concern for their potential to
cause development of resistance in pest species by various sectors of
pest control science.

In 1982 it was reported from this laboratory (Ghiasuddin et al.,
1980) that the Ca-ATPase from the central nervous system of the DDT
resistant VPIDLS strain of the German cockroach Blattella germanica
is less sensitive to DDT as compared to that from the susceptible
counterpart, CSMA strain. The system from the resistant strain
appeared to be much less responsive to changes in externally applied
Ca^{2+} concentration than was the susceptible counterpart (Fig. 6).

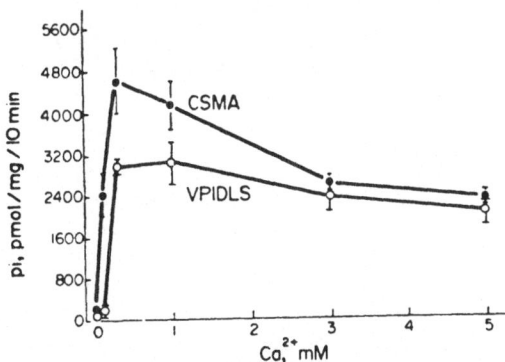

Fig. 6. Effect of Ca^{2+} on ATPase activity of the membrane fraction
from DDT-susceptible (CSMA) and DDT-resistant strains
(VPIDLS) of German cockroach under standard assay conditions
at $30^{\circ}C$ as given in materials. Ten μg equivalent protein per
assay tube was employed. The assay buffer contained 60 mM
Na^{+}, 60 mM K^{+}, pH 7.3 previously adjusted with 30 mM Tris-
HCl. The vertical lines indicate standard error for each
point. Averages of three independent experiments. (From
Ghiasuddin et al., 1981).

In view of the recent realization that "Ca-ATPase" is likely to
be identical to the Na-Ca protein kinase-phosphatase system (PKP), we
have recently reexamined the above phenomenon (Rashatwar and
Matsumura, 1985). When the nerve Na-Ca PKP activity in the resistant
and susceptible cockroaches was assayed it became apparent that the
Ca^{2+} response of the PKP from the brain synaptosome preparation from
the resistant B. germanica is much less than that from the susceptible strain (Table 1).

By this time adequate methodologies have been developed to assay
both direct $^{45}Ca^{2+}$ transport and Na^{+} channel activity. When these
nerve preparations were examined through these assay methods for
their requirements for Ca^{2+} it became apparent that the resistant
system is far less responsive to Ca^{2+} concentration changes that the
corresponding system from the susceptible cockroaches. Not only
that, the rate of ^{22}Na uptake by the brain synaptosomes was clearly
stimulated at low Ca^{2+} concentrations only in the susceptible strain
and not in the resistant strain (Fig. 7). This portion of $^{22}Na^{+}$
uptake is antagonized by tetrodotoxin, and, therefore, likely to
represent the sodium channel activity. When these insects were
tested in vivo for toxicity against agents which are known to interfere with Ca^{2+} regulation, it has become evident that the resistant
strain shows strong cross-resistance to A23187 and theophylline in
addition to their cross-resistance to poisons known to interact with
the Na^{+} inactivation mechanism (e.g., grayanotoxin I and veratrin)

Table 1. DDT inhibition and calmodulin stimulation of Na^+-Ca^+ protein
          kinase phosphatase (=Ca ATPase) activity of the synaptosome
          preparations from DDT-susceptible (CSMA) and resistant
          strain (VPIDLS) of the German cockroach

| | (Pi) Production (pmole/mg/10 min)[b] | |
Treatment[a]	CSMA	VPIDLS
Basal (EGTA)	306 ± 22	132 ± 36
Basal + DDT	114 ± 18	78 ± 27
+ Calcium	432 ± 43	245 ± 27
+ Calcium + Calmodulin	534 ± 25	298 ± 30

+ Calcium
  + Calmodulin
  + DDT[a]
	245 ± 16	124 ± 29

[a]Concentrations were: DDT 10^{-5}M, calmodulin 344 nM, and calcium
10^{-6}M (given as a Ca^{2+}-EGTA complex).
[b]Data expressed as mean ± SE of 3 to 6 determinations.

From Rashatwar and Matsumura, 1985.

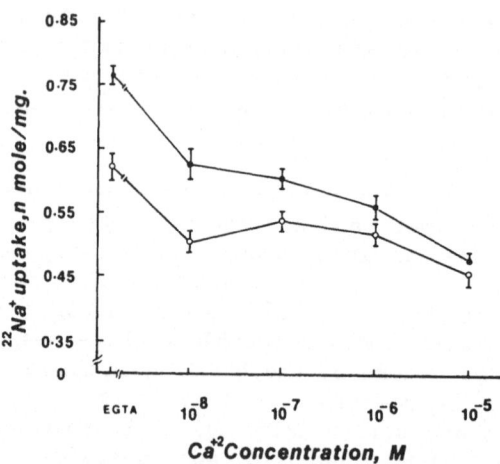

Fig. 7. Effect of changes in Ca^{2+} concentration and DDT 10^{-5}M on
         the level of $^{22}Na^+$ uptake activity of the synaptosome
         preparations from DDT-susceptible (CSMA) and resistant
         strain (VPIDLS) of German cockroaches at room temperature
         under standard assay condition, (●) CSMA, (o) VPIDLS. (From
         Rashatwar and Matsumura, 1985).

(Table 2). The results indicate that the insects can withstand increases in intracellular Ca^{2+} and such a mechanism is somehow rendering them the ability to resist against poisons affecting the Na^+ channel.

Overall Scheme of Calcium Participation in the Phosphorylation

The above study has shown that a point mutation occured in the resistant nervous system to reduce Ca^{2+} sensitivity which had the overall effects of making Na-Ca PKP and Na-channel systems less sensitive to DDT. To explain the phenomenon I have developed a hypothetical scheme by which these systems are operated.

$$Pr^* \xrightarrow[\substack{\text{high } K^+, Mg^{2+} \\ \text{low } Na^+}]{(1) \text{ ATP}} Pr^{**}-P \xrightarrow[\text{low } Ca^{2+}, \text{ Ca M}]{(2) \text{ DDT, pyrethroid}} Pr^{**}-Ca-P \xrightarrow[]{(3) \text{ low } K^+, \text{ high } Na^+, Ca^{2+}}$$

Where Pr is either Na/Ca exchanging protein, Na^+-channel protein or other calcium requiring proteins. Ca M designates calmodulin.

The phosphorylated protein (Pr-P**) is expected to have a higher affinity to Ca^{2+} (Blaustein, 1977) than the unphosphorylated protein (Pr*). Since DDT's action can be antagonized or overcome by addition of excess Ca^{2+} added externally (Matsumura and Narahashi, 1971) and calmodulin (Hagemann, 1982; Rashatwar and Matsumura, 1984), it may be reasoned that step (2) is the critical step where DDT's action becomes detrimental to the overall process. The lack of Ca^{2+} response in the Na-channel and PKP system from the Kdr-type resistant insect nerves could be interpreted to mean that a change took place in their nerve to eliminate or reduce the degree of reliance on calcium. In such a system a small amount of calcium required for operation of Na-channel for instance could be already tied up with the protein or made available through an entirely different mechanism, and, therefore, the system becomes more independent from the action of chemicals which affect the availability of calcium. The fact that the resistant insects show cross-resistance to agents which cause hypocalcemic (TPZ, EGTA, etc.) as well as those which promote hypercalcemic conditions (theophylline, A23187) supports the above view.

The proposed meaning of calcium binding to the active membrane protein is stabilization of the channel operation in the case of sodium channel, and uptake, sequestration and eventual transport of Ca^{2+} in the case of Na/Ca exchange. How the resistant insects with low calcium responding membrane protein systems carry out normal nerve functions is not clear at all. It would take much more work to elucidate underlying mechanisms. On the other hand, the fact that these insects exhibit such resistance to pyrethroids (Scott and

Table 2.   Susceptibility levels (24 hr LD_{50}) of resistant and susceptible strains of German cockroach against various agents[a]

	Cockroach Strains				
	CSMA	VT	Ratio[b] (Resistance)	VPIDLS	Ratio[b] (Resistance)
Insecticides					
DDT	60	121	(2.0)	320	(5.3)
Diazinon	0.25	0.30	(1.2)	---	-----
Nicotine	37	36	(0.97)	---	-----
Carbaryl	18[e]	14[e]	(0.78)	21[e]	(1.2)
Agents affecting Na$^+$, K$^+$ channel					
Grayanotoxin I	18	48	(2.7)	80	(4.4)
Valinomycin	0.44	0.50	(1.1)	1.8	(4.0)
Aconitine	6.6	10	(1.5)	16	(2.4)
Veratrin	3.4	7.1	(2.1)	11	(3.2)
Calcium[f] modulator					
Calmidazolium	44	39	(0.88)	108	(2.5)
TPZ[c]	64	96	(1.5)	179	(2.8)
Chlorpromazine	24	17	(0.71)	92	(3.9)
A23187[d]	1.3	6.4	(4.8)	13	(9.4)
Gramicidin D	3.1	5.4	(1.8)	8.6	(2.8)
EGTA[c]	91	206	(2.3)	291	(3.2)
Lathanum	65	66	(1.1)	73	(1.1

[a]Data are expressed in terms of dose needed to kill 50% of
population. All roaches were treated with 30 μg/roach of
piperonyl butoxide (topical) prior to the tests.
[b]Resistant ratio between LD_{50} of resistant (VT or VPIDLS) divided
by LD_{50} of susceptible (CSMA) strain. The value less than 1.0
indicates that the susceptible strain is more resistant.
[c]TPZ (=trifluoroperazine) EGTA (Ethylene glycol-bis (p-
aminoethyl) ether N, N, N, N-tetraacetic acid).
[d]Calcium ionophore.
[e]LT_{50} in hours with piperonyl butoxide.
[f]Other compounds to which VT did not show any significant cross-
resistance were: D600, nitrindipine, verapamil and PCMPS.

From Rashatwar and Matsumura, 1985.

Matsumura, 1983) clearly indicates the importance of calcium mediating processes in the action mechanism of these insecticides.

   A question may be raised at this time how widely this change in
calcium binding in Kdr resistant insects affects other calcium utilizing systems. Two systems which can be eliminated from this consideration are calmodulin and acetylcholine receptor. This conclu-

sion is based on the observation that the resistant insects do not exhibit any significant cross-resistance to calmidazolium, a specific calmodulin inhibitor, and to nicotine (Rashatwar and Matsumura, unpublished data). It may be concluded, therefore, that the change in calcium affinity does not involve all calcium requiring systems. It could be that Na/Ca exchange system is constructed in a very similar fashion as Na^+-channel, at least in terms of Ca^{2+} binding mechanisms. Both systems have high and low Ca^{2+} sites in addition to Na^+ binding site. They show rather low temperature quotients and sensitivities to lipophilic agents such as veratridine, DDT and pyrethroids.

DISCUSSION

In conclusion, enough evidence has accumulated to show that DDT and pyrethroids affect calcium regulatory mechanisms in the nervous system. Particularly sensitive to these agents are Na^+ channel and Na-Ca protein kinase-phosphatase system (formerly called Ca-ATPase). The theory of involvement of calcium regulatory systems in the action of DDT and pyrethroids is supported by the observation that (a) addition of excess Ca^{2+} in isolated nerve preparations antagonize their actions, (b) DDT's action in vitro on Na-Ca PKP may also be eliminated by the addition of excess calmodulin, (c) Kdr-type resistant insects have Na^+-channel and Na/Ca exchange systems with very low Ca^{2+} response, and (d) they show in vivo clear-cut cross-resistance to A23187 (calcium ionophore), theophylline and EGTA, the agents known to change cellular Ca^{2+} concentrations.

Certainly much more knowledge would be required to pinpoint the precise mechanism of action of these insecticides at the channel or effector level. However, sound progress has been made in that direction. The progress in this field coincides with the vast advancement in neurobiochemistry, particularly in the areas related to calcium regulation and roles of phosphorylation of membrane proteins in functional modulation.

ACKNOWLEDGEMENTS

Supported in part by the Michigan Agricultural Experiment Station, and research grant ESO1963 from the National Institute of Environmental Health Science, Research Triangle Park, North Carolina. The unpublished experimental works presented here are those produced by Drs. J. Marshall Clark, S. Rashatwar and J.G. Scott.

REFERENCES

Blaustein, M.P., 1977, Effects of internal and external cations and
    of ATP on sodium-calcium exchange in squid axons, Biophys. J.,
    20:79-111.
Blaustein, M.P., 1974, The interrelationship between sodium and cal-
    cium fluxes across cell membranes, Rev. Physiol. Biochem.
    Pharmacol., 70:33-82.
Bratkowski, T.A., and Matsumura, F., 1972, Properties of a brain
    adenosine triphosphatase sensitive to DDT, J. Econ. Entomol.,
    65:1238-1245.
Clark, J.M., and Matsumura, F., 1982, Two different types of inhibi-
    tory effects of pyrethroids on nerve Ca- and Ca+ Mg-ATPase
    activity in the squid, Loligo pealei, Pestic. Biochem. Physiol.,
    18:180-190.
Cutkomp., L.K.,, Koch, R.B., and Desaiah, D., 1982, Inhibition of
    ATPases by chlorinated hydrocarbons, in: "Insecticide Mode of
    Action,"J.R. Coats, ed., Academic Press, New York.
DiPolo, R., 1973, Calcium efflux from internally dialyzed squid
    giant axons, J. Gen. Physiol., 62: 575-589.
DiPolo, R., and Beauge, L., 1983, The calcium pump and sodium cal-
    cium exchange in squid axons, Ann. Rev. Physiol., 45:313-324.
Doherty, J.D., and Matsumura, F., 1974, DDT effect on ^{32}P incorpora-
    tion from gamma-labeled ATP into proteins from lobster nerve, J.
    Neurochem., 22:765-772.
Georghiou, G.P., and Saito, T., (eds), 1982, "Pest Resistance to
    Pesticides", Plenum Press, New York.
Ghiasuddin, S.M., Kadous, A.A., and Matsumura, F., 1980, Reduced
    sensitivity of a Ca-ATPase in the DDT-resistant strains of the
    German cockroach, Comp. Biochem. Physiol., 68C:15-20.
Ghiasuddin, S.M. and Matsumura, F., 1979, DDT inhibition of Ca-
    ATPase of the peripheral nerve of the American lobster, Pestic.
    Biochem. Physiol., 10:151-161.
Gonda, O., Klauszner, A., and Avidor, Y., 1959, Effect of 1,1,1-
    trichloro-2,2-di(p-chlorophenyl)ethane (DDT) and related com-
    pounds on the adenosine triphosphate-phosphate exchange cata-
    lyzed by a particular fraction from the mosquito, Biochem. J.,
    73:583-587.
Gordon, H.T., and Welsh, J.H., 1948, The role of ions in axon sur-
    face reactions to toxic organic compounds, J. Cellular Comp.
    Physiol., 31:395-420.
Hagemann, J., 1982, Inhibition of calmodulin-stimulated cyclic
    nucleotide phosphodiesterase by the insecticide DDT, Fed. Eur.
    Bioch. Soc. Lett., 143:52-54.
Koch, R.B., 1969, Chlorinated hydrocarbon insecticides: inhibition
    of rat brain ATPase activities, J. Neurochem., 16:269-271.
Kearns, C.W., 1956, The mode of action of insecticides, Ann. Rev.
    Entomol., 1:123-148.
Matsumura, F., 1970, DDT action and adenosine triphosphate related
    systems, Science, 169:1343.

Matsumura, F., and Clark, J.M., 1980, ATPases in the axon-rich membrane preparation from the retinal nerve of the squid, Loligo pealei., Comp. Biochem. Physio., 66B:23-32.

Matsumura, F., and Clark, J.M., 1982, ATP-utilizing systems in the squid axons: a review on the biochemical aspects of ion-transport, Progr. Neurobiol., 18:231-255.

Matsumura, F., Bratkowski, T.A., and Patil, K.C., 1969, DDT: inhibition of an ATPase in the rat brain, Bull. Environ. Contam. Toxicol., 4:262-270.

Matsumura, F., and Ghiasuddin, S.M., 1979, Characteristics of DDT-sensitive Ca-ATPase in the axonic membrane, in: "Neurotoxicology of Insecticides and Pheromones," T. Narahashi, ed., Plenum Press, New York.

Matsumura, F., and Patil, K.C., 1969, Adenosine triphosphatase sensitivity to DDT in synapses of rat brain, Science, 166:121-122.

Mullins, L.J. and Brinley, F.J., Jr., 1967, Some factors influencing sodium extrusion by internally dialyzed squid axons, J. Gen. Physiol., 50:2333-2355.

Narahashi, T., 1979, Nerve membrane ionic channels as the target site of insecticides, in: "Neurotoxicology of Insecticides and Pheromones," T. Narahashi, ed., Plenum Press, New York.

Narahashi, T., 1976, Effects of insecticides on nervous conduction and synaptic transmission, in: "Insecticide Biochemistry and Physiology," C.F. Wilkinson, ed., Plenum Press, New York.

O'Brien, R.D., 1976, "Insecticides: Action and Metabolism," Academic Press, New York.

Rashatwar, S., and Matsumura, F., 1985, Reduced calcium sensitivity of the sodium channel and the Na/Ca exchange system in the Kdr type, DDT and pyrethroid resistant German cockroaches, Blattella germanica, Comp. Biochem. Physiol., In Press.

Scott, J.G., and Matsumura, F., 1983, Evidence for two types of toxic actions of pyrethroids on susceptible and DDT-resistant German cockroaches, Pestic. Biochem. Physiol., 19:141-150.

Scott, J.G. and Matsumura, F., 1981, Characteristics of DDT induced case of cross-resistance to permethrin in Blattella germanica, Pestic. Biochem. Physiol., 16:21-27.

Woolley, D.E., 1982, Neurotoxicity of DDT and possible mechanism of action, in: "Mechanisms of Action of Neurotoxic Substances," K.N. Prasad and A. Vernadakis, eds., Raven Press, New York.

ACTION OF PYRETHROIDS ON CA^{2+}-STIMULATED ATP HYDROLYZING ACTIVITIES: PROTEIN PHOSPHORYLATION-DEPHOSPHORYLATION EVENTS IN INSECT BRAIN FRACTIONS

J. Marshall Clark

Department of Entomology, University of Massachusetts
Amherst, Massachusetts 01003

INTRODUCTION

Synaptosomal and microsomal preparations of cockroach brain elicited Ca^{2+}- stimulated ATP hydrolyzing activities. Two Ca^{2+}- stimulated ATP hydrolyzing systems (i.e., Na-Ca ATP hydrolysis and Ca+Mg ATP hydrolysis) were found to be highly sensitive to a variety of pyrethroid insecticides. Overall, the synaptic preparation appeared to be the more sensitive site of pyrethroid action. This was particularly true in the case of nonmitochondrial Ca+Mg ATP hydrolysis in disrupted synaptosomes which was the most sensitive activity examined.

Allethrin, a type I pyrethroid, elicited maximal inhibition on Na-Ca ATP hydrolysis, whereas cypermethrin, a type II pyrethroid, mainly inhibited Ca+Mg ATP hydrolysis. This selectivity of action was not complete however in that both types of pyrethroids were clearly inhibitory to both ATP hydrolyzing systems.

These _in vitro_ findings have been substantiated by _in vivo_ experiments where permethrin, a non-alpha cyano analog of cypermethrin, was administered to live roaches. All Ca^{2+}-stimulated ATP hydrolyzing activities were reduced by this treatment. More importantly, this _in vivo_ inhibition occurred in the presence of a similar amount of bound insecticide which caused a substantial amount of inhibition _in vitro_.

REVIEW OF RELATED WORK

One of the most obvious and characteristic aspects of DDT and pyrethrin poisoning is the appearance of repetitive electric dis-

189

charges recorded electrophysiologically from nerve membranes. The
failure of a number of highly synthetic pyrethroids and some DDT
analogs to elicit identical responses indicates that perhaps there
are other aspects involved in this process in addition to the axonal
sodium gating phenomena (Nishimura and Narahashi, 1978; Narahashi,
1979; Vijverberg and Van der Bercken, 1979; Gammon et al., 1981).

Regulation of cytoplasmic concentration of calcium is particularly important in the nervous system as the level of free calcium
affects both axonal conduction and synaptic transmission (Mullins,
1981). Frankenhaeuser and Hodgkin (1957) initially reported that a
reduction of external calcium on nerve axons resulted in spontaneous
electrical oscillations and nerve instability. This repetitive
activity mimicked the repetitive discharges seen during DDT
poisoning. Similarly, neurotransmitter release from presynaptic
nerve terminals has been shown to be dependent upon intraterminal
Ca^{2+} concentration (Rahamimoff et al., 1976; Llinas et al., 1982).

The biological ramifications which occur upon Ca^{2+} entry during
a single nerve discharge impose at least two necessary restrictions
on Ca^{2+} regulation in excitable cells. First, intracellular resting
Ca^{2+} concentrations must be maintained at exceedingly low levels
(e.g., 10^{-7} M or less) for the cell to detect a Ca^{2+} signal produced
during bioelectrical activity. Secondly, Ca^{2+} entry must be highly
interfaced with Ca^{2+} extrusion or sequestration systems in order to
maintain Ca^{2+} homeostasis (Mullins, 1981). For these reasons, an
arrangement of millimolar quantities of calcium on the outside and
submicromolar quantities on the inside (i.e., free ionized Ca^{2+}) has
been shown to be an absolute requirement for nerve excitability
(Tasaki, 1974).

At least four mechanisms which extrude or sequester Ca^{2+} in
nerve cells have been implicated in the control of intracellular Ca^{2+}
concentration; the plasma membrane, endoplasmic reticulum, intrasyn-
aptic vesicles and mitochondria (Gill et al., 1984). Of these,
plasma membrane and endoplasmic reticulum play pivotal roles in
regulation of cytosolic Ca^{2+}. The plasma membrane functions in the
ultimate removal of cytosolic Ca^{2+} from the nerve cell whereas endoplasmic reticulum with its huge surface area functions as a major
internal system for sequestration of cytosolic Ca^{2+}.

There is ample evidence supporting the coexistence of two Ca^{2+}
efflux mechanisms (i.e., (ATP + Mg^{2+})-dependent Ca^{2+} pump and Na/Ca
exchange) operating in neural plasma membranes (DiPolo et al., 1979;
Gill et al., 1981; 1984) and of an ATP-dependent Ca^{2+} uptake into
endoplasmic reticulum which shares many similarities with the Ca^{2+}
pump in sarcoplasmic reticulum (Blaustein et al., 1978a; McGraw et
al., 1980). Furthermore, it is now understood that Na/Ca exchange in
a variety of excitable cells is promoted by ATP. This process has
been detailed and found to be a cyclic mechanism in which a protein

kinase activates the exchange in a phosphorylation step and a phosphatase deactivates it in a dephosphorylation step (Caroni and
Carafoli, 1983). This ATP-promoted aspect of Na/Ca exchange has been
shown to be saturated at 1.0 μM ATP which is indicative of a very high
affinity for ATP (Reinlib et al., 1981). Therefore, a unifying
concept of Ca^{2+} homeostasis within nerve tissues is its utilization
of ATP either directly as an energy source (e.g., adenosine triphosphatases) or indirectly as a modifier or regulator (e.g., protein
kinase-phosphatase).

The presence of ion-stimulated ATP hydrolyzing activities in a
variety of excitable tissues has been extensively documented (Abood
and Gerad, 1954; Skou, 1957) and their sensitivity towards DDT and
pyrethroids is well established (Matsumura and Patil, 1969; Doherty
and Matsumura, 1975; Janicki and Kinter, 1971; Desaiah et al., 1974,
Huddard, 1974; Schneider, 1975; Miller et al., 1976). Furthermore, a
number of observations has indicated a high degree of sensitivity of
Ca^{2+}-stimulated ATP hydrolyzing activities to other insecticidal
compounds (Ghiasuddin and Matsumura, 1979; Yamaguchi et al., 1979).

Matsumura and Clark (1980) have characterized two types of
Ca^{2+}-stimulated ATP hydrolyzing activities from the squid nervous
system. One is Ca+Mg ATP hydrolysis which is temperature and aging
sensitive and requires the simultaneous presence of Mg^{2+} and K^+ for
optimal Ca^{2+} stimulation. The second is Na-Ca ATP hydrolysis which
is relatively temperature insensitive, stable and resistant to aging,
and requires the simultaneous presence of $Na^+(Li^+)$ and K^+ but not
Mg^{2+}, for optimal Ca^{2+} stimulation. Both activities are ouabain-
insensitive and show high affinity for ATP (Matsumura and Clark,
1982). Furthermore, ATP hydrolysis apparently occures at least in
part via a membrane phosphorylation - dephosphorylation process
which is highly dependent on calcium concentration (Matsumura and
Clark, 1980). Using these criteria, Ca^{2+}-stimulated ATP hydrolysis
activities have been found to be highly sensitive to the action of by
pyrethroid insecticides. It was shown subsequently that pyrethroids
such as pyrethrin and allethrin are most inhibitory to Na-Ca ATP
hydrolysis whereas cypermethrin and decamethrin preferentially affect
Ca+Mg ATP hydrolysis (Clark and Matsumura, 1982). Indeed, the
actions of various pyrethroid compounds have been separated into
remarkably similar categories based on both their electrophysiological (Gammon et al., 1981) and symptomological (Aldridge, 1980)
responses.

In view of the growing utilization of pyrethroids, it appears
prudent to study the toxicological significance of this newly discovered property in organisms targeted for control with these chemicals.

EXPERIMENTAL

Preparation of Insect Brain Fractions

Adult male cockroaches, Periplaneta americana, were used exclusively in this study. Roach brains were obtained and fractionated in the manner described by Telford (1968). A 105,000 x g pellet, free of mitochondrial contamination, was obtained by differential ultracentrifugation and was utilized as the microsomal tissue fraction.

A 20,000 x g pellet (crude mitochondrial fraction) which was obtained previously by differential ultracentifugation was fractionated further by a discontinous sucrose gradient (Telford, 1968). The material at the 1.2 M –1.5 M interface was collected by aspiration and served as the synaptosomal tissue fraction.

Determination of ATP Hydrolysis

The hydrolysis of gamma ^{32}P-labeled adenosine triphosphate (γ-^{32}P ATP) (New England Nuclear, Boston, MA) and the subsequent production of radiolabeled inorganic phosphate (^{32}Pi) was used as an indicator of ATP hydrolyzing activity in all assays. The assay methods were exactly as previously described (Matsumura, 1977; Matsumura and Clark, 1980; Clark and Matsumura, 1982).

As originally discussed by Matsumura and Clark (1980) and reviewed recently (1982), the assay methods used here are not able to define a specific enzyme system or protein but rather were designed to determine specific ion requirements resulting in increased ATP hydrolysis, a portion of which might be applied to ion translocations. The most well understood ATP-stimulated ion transporting systems are adenosine triphosphatases (i.e., ATPases) such as the sodium or calcium pumps. The assay methods employed here are not able to differentiate a phosphorylation–dephosphorylation event such as occurs during a coupled protein kinase–phosphoprotein phosphatase reaction from that which would occur during an adenosine triphosphatase reaction. In this paper, we have adopted the more economical term of "ATP hydrolysis" to encompass all ATP hydrolyzing activities and should not to be restricted to ATP-dependent pump activities in the strict biochemical definition of ATPase enzymes.

All ATP hydrolyzing activities were stable to storage on ice (4°C) except Ca+Mg ATP hydrolysis which was assayed only with freshly prepared tissue fractions and used immediately. Protein was determined by the method of Lowry et al. (1951), and all assays were performed at a standard protein concentration of 10 μg protein per assay tube. Approximately 3.0×10^{5} cpm of γ-^{32}P ATP isotope were used per assay tube.

Inhibition of ATP Hydrolyzing Activities

All experimental agents were prepared in buffer solutions or in organic solvents at stock concentrations adjusted to deliver the final assay concentration (1 ml) in a 10 μl aliquot. All agents, unless stated otherwise in the text, were equilibrated with the tissue-buffer mixture for a preincubation period of 10 min at 25°C prior to the addition of γ-^{32}P ATP. Whenever organic solvents were utilized, solvent controls were included.

Determination of Permethrin Binding to Insect Brain Fractions

In Vivo Studies. In order to determine the amount of permethrin bound to brain tissue in severely poisoned roaches, twenty live male roaches were injected intraperitoneally at an LD_{95} dose of radio-labeled ^{14}C-permethrin (1 μg/roach, Chadwick, 1979). The amount of ^{14}C-permethrin bound to whole brains, microsomal, and synaptosomal fractions was determined by liquid scintillation spectrometry. Cis-permethrin (^{14}C-carbonyl) was utilized throughout this study in order to minimize metabolic conversion and had a specific activity of 58.2 mCi/mmole. Twenty roach brains contained approximately 15.3 mg protein. Microsomal and synaptosomal fractions from this amount of brain contained 1020 μg and 420 μg protein, respectively.

A collaborative study was attempted to determine the affects of permethrin binding utilizing unlabeled cis-permethrin at the same LD_{95} dose. Roaches were treated exactly as outlined above except now permethrin-containing brain fractions were utilized to measure the specific activities of various Ca^{2+}-stimulated ATP hydrolyzing activities under standard assay conditions.

In Vitro Studies. In these experiments, an I_{50} dose of cis-permethrin (^{14}C-carbonyl), calculated for nonmitochondrial Ca+Mg ATP hydrolysis (i.e., 3.8×10^{-7}M for microsomes, 2.3×10^{-8}M for synaptosomes, see Fig. 5), was added during the 10 min preincubation period of a standard ATP hydrolysis assay. The assay was terminated by quick cooling and tissue fractions collected by centrifugation using the method of Matsumura and Hayashi (1969). The amount of ^{14}C-permethrin bound to microsomal and synaptosomal fractions was determined by liquid scintillation spectrometry. Unlabeled tris-ATP was used as a substrate in order not to interfere with ^{14}C-permethrin determination.

Chemicals

Pyrethroid insecticides used in this study are as follows: (+)-trans allethrin [3-allyl-2-methyl-4-oxocyclopent-2-enyl-(+)-trans-chrysanthemate of (+)-allethrolone], permethrin [3-phenoxybenzyl(+)-cis, trans-(2,2-dichlorovinyl)-2,2-dimethyl-cyclopropanecarboxylate], cis-permethrin (^{14}C-carbonyl from FMC, Agricultural Chemicals Divi-

sion, Middleport, NY), and cypermethrin [(+)-α-cyano-3-phenoxybenzyl
(+)-cis,trans-3-(2,2-dichlorovinyl)-2,2-dimethylcyclopropanecarboxy-
late] from Shell Bioscience Laboratory, Sittingbourne, Kent, U.K..
The following pharmacological active agents were obtained from
various sources; carbonylcyanide-p-trifluoromethoxy-phenylhydrazone
(FCCP) from Boehringer Mannheim GmbH, calcium ionophore A23187 from
Eli Lilly Company, valinomycin from Aldrich Chemical Company,
tiapamil from F. Hoffmann-LaRoche Company, trifluoperazine-2HCl(TFP)
from Smith, Kline & French Corporation, dithiobis(succinimidyl propionate) (DSP) from Tridom Chemical Inc., sodium orthovanadate from
Fisher Scientific Company, and monensin from Calbiochem-Behring Corp.
The remaining chemicals were all received from Sigma Chemical
Company; dithiotheitol (DTT), ethylenediamine tetraacetic acid
(EDTA), imadazole, ethyleneglycol bis(β-aminoethylether)-N,N'-tetra-
acetic acid (EGTA), tris(hydroxymethyl)aminomethane-ATP (tris-ATP),
ouabain, lanthanum chloride (LaCl$_3$), ruthenium red, colchicine,
cytochalasin B, tetrodotoxin (TTX), mersalyl acid, trypsin, trypsin
inhibitor (soybean, type 1-S), tetracaine, atractyloside, veratridine, pentylenetetrazole (PTZ), saponin (S), and the mitochondrial
poisons (M)= (oligomycin, sodium azide, and 2,4, dinitrophenol
(DNP)).

Characterization of Insect Brain ATP Hydrolyzing Activities

Table 1 summarizes the buffer compositions used to characterize
Ca^{2+}-stimulated ATP hydrolyzing activities in the insect brain. As
previously described by Matsumura and Clark (1980, 1982), these

Table 1.   Composition of standard buffer solutions[a] used to
characterize each ATP hydrolyzing system

ATP hydrolysis buffers	NaCl	KCl	MgCl$_2$	CaCl$_2$[b]	Imadazole	EGTA	EDTA	Ouabain
					Concentration, mM			
Basal Na-Ca	160	160	0	0	30	0	1	0.1
1 mM Na-Ca	160	160	0	1	30	0	1	0.1
50 uM Na-Ca	160	160	0	0.05	30	0	1	0.1
0.5 uM Na-Ca	160	160	0	5x10^{-4}	30	0	1	0.1
Basal Mg	0	160	10	0	30	0.5	0	0.1
Ca+Mg	0	160	10	0.01	30	0.5	0	0.1
(Na)Ca+Mg	160	160	10	0.01	30	0.5	0	0.1
(Li)Ca+Mg	(Li=160)	160	10	0.01	30	0.5	0	0.1

[a]All buffer solutions were adjusted to pH 7.3 using HCl. All water utilized in
experimentation was of reagent grade quality (i.e., Type 1).
[b]Free ionized Ca^{2+} concentrations were adjusted using CaEDTA or CaEGTA buffer
systems at levels previously described by Portzehl et al. (1964).

activities appear to separate into two groups; those in which Ca^{2+} stimulation requires Mg^{2+}, and are highly temperature sensitive (i.e., Ca+Mg ATP hydrolysis) and those which are Mg^{2+} independent, require $Na^+(Li^+)$, and are relatively temperature insensitive (i.e., Na-Ca ATP hydrolysis). Table 2 shows that similar activities are found in both microsomal and synaptosomal tissue fractions of the insect brain. Na-Ca ATP hydrolysis is consistently higher in the microsomal preparation as compared to the synaptosomal fraction. Ca+Mg ATP hydrolysis is approximately equal in the two preparations as judged by total activity. It is also noteworthy that nonmitochondrial Ca+Mg ATP hydrolysis is greatly reduced by the presence of Na^+ or Li^+. A similar phenomenon has been reported by Blaustein et al. (1978b) concerning the ATP-dependent nonmitochondrial Ca^{2+} transport mechanism in mammalian synaptosomes.

The integrity of the synaptosomal fraction has been judged by using a naturally occuring surfactant, saponin. Saponin disruption is a less destructive means of rupturing intact membrane systems than the more often used method of osmotic shock. It is believed that saponin interferes principally with cholesterol-rich membranes (e.g., plasmalemma) leaving cholesterol-poor membranes (e.g., inner mitochondrial membranes or endoplasmic reticulum) intact. Saponin disruption of insect brain synaptosomes increases basal Mg-ATP hydrolysis. Also the heretofore absent Ca^{2+} stimulation, which is apparent in undisrupted microsomal fractions, is now measurable but is evident only in the combined presence of saponin disruption and mitochondrial poisons; azide, DNP and oligomycin. This indicates that the Ca^{2+} binding sites resulting in Ca^{2+} stimulation of Ca+Mg ATP hydrolysis are internally situated either on the inside of the plasmalemma or associated with internal membrane systems such as endoplasmic reticulum.

Mitochondrial versus nonmitochondrial ATP hydrolysis in disrupted synaptosomes can also be separated using a variety of selective inhibitors. The results of these experiments are outlined in Table 3. Ruthenium red at a low concentration ($10^{-8}M$) has been implicated as a very specific inhibitor of Ca^{2+} sequestering by mitochondria (Carafoli and Crompton, 1978). Similar levels of ruthenium red decreased the apparent mitochondrial Mg-ATP hydrolysis in the insect brain, but had no inibitory effect on nonmitochondrial Ca+Mg ATP hydrolysis. FCCP, an extremely potent uncoupling agent of oxidative phosphorylation, is also inhibitory to mitochondrial activity. The mitochondrial ATP-ADP carrier inhibitor, atractyloside (Lehninger, 1977), inhibits mitochondrial activity by some 80%.

Blaustein et al. (1978a) reported the proteolytic enzyme, trypsin, can also be used to delineate mitochondrial versus nonmitochondrial activities in disrupted synaptosomes. Proteolytic digestion was accomplished by incubating saponin disrupted synaptosomes with trypsin (1 mg/ml) for 5 min at 30°C. Trypsin digestion was

Table 2. Distribution of ATP hydrolyzing activities in microsomal and synaptosomal fractions of the insect brain

ATP hydrolysis buffers	Tissue[b]	Pre-Incubation[c] S	Pre-Incubation[c] MP	ATP hydrolyzing activities+S.E.[d] Total[e]	ATP hydrolyzing activities+S.E.[d] Ca^{2+}-stimulated[f]	ATP hydrolyzing activities+S.E.[d] Mitochondrial[g]	ATP hydrolyzing activities+S.E.[d] Non-Mitochondrial[h]
				(pmoles Pi/mg protein/10 min)			
Basal Na-Ca	micro	–	–	48+17			
1 mM Na-Ca	micro	–	–	3320+210	3272+220		
50 uM Na-Ca	micro	–	–	2016+234	1968+232		
0.5 uM Na-Ca	micro	–	–	227+35	179+33		
Basal Na-Ca	syn	–	–	121+3			
1 mM Na-Ca	syn	–	–	1804+180	1683+160		
50 uM Na-Ca	syn	–	–	1286+210	1165+240		
0.5 uM Na-Ca	syn	–	–	272+10	151+19		
Basal Mg	micro	–	–	673+25		76+13	
Basal Mg	micro	–	+	597+51			597+51
Ca+Mg	micro	–	–	1053+98	380+42	134+9	
Ca+Mg	micro	–	+	919+84	322+60[h]		919+84
(Na)Ca+Mg	micro	–	+	683+42	86+30		
(Li)Ca+Mg	micro	–	+	634+64	37+12		
Basal Mg	syn	–	–	458+25			
Basal Mg	syn	+	–	1046+89		460+18	
Basal Mg	syn	+	+	586+53			586+53
Ca+Mg	syn	–	–	429+14	0		
Ca+Mg	syn	+	–	1042+74	0	315+14	
Ca+Mg	syn	+	+	727+47	141+19[h]		727+47

[a] ATP hydrolysis refers to ATP hydrolyzing activity selectively stimulated by ionic buffers described in Table 1.
[b] Micro (microsomal) and syn (synaptosomal), respectively.
[c] Intact synaptosomal tissue fractions were disrupted by pretreatment with saponin (S) at a concentration of 250 ug/ml for 30 min at $4°C$. Mitochondrial poisons (MP) consist of a mixture of 0.1 mM azide, 0.1 mM DNP and 0.7 ug/ml oligomycin (Blaustein et al., 1978a).
[d] Specific activity of each ATP hydrolyzing system is reported as the mean (pmole Pi/mg protein/10 min)+standard error of at least six separate experiments of two replicates each.
[e] Total ATP hydrolysis is defined as the ATP hydrolyzing activity apparent in the presence of a full ion complement for each buffer solution.
[f] Ca^{2+}-stimulated ATP hydrolysis refers to the ATP hydrolyzing activity determined as the difference between Total activity (plus Ca^{2+}) and Basal activity (minus Ca^{2+}).
[g] Mitochondrial ATP hydrolysis refers to the difference in ATP hydrolyzing activity in the presence and absence of mitochondrial poisons (see footnote c above).
[h] Nonmitochondrial ATP hydrolysis refers to the ATP hydrolyzing activity remaining in the presence of mitochondrial poisons.

terminated by addition of trypsin inhibitor (1.5 mg/ml). As seen in Table 3, trypsin digestion reduces the nonmitochondrial Ca+Mg ATP hydrolysis but has no effect on mitochondrial Mg-ATP hydrolysis. Trypsin plus trypsin inhibitor had no effect on the mitochondria activity but did effectively protect a portion of the nonmitochon-

Table 3.   Influence of cellular inhibitors on mitochondrial Mg–ATP
           hydrolysis and nonmitochondrial Ca+Mg ATP hydrolysis from
           disrupted synaptosomes[a] of the insect brain

Additions	Mitochondrial Mg–ATP Hydrolysis[b]	Nonmitochondrial Ca+Mg ATP Hydrolysis[c]
	Percent of control ± S.E.[d]	
Ruthenium red (10^{-8}M)	51.4±2.7	+100[f]
Ruthenium red (10^{-7}M)	6.8±4.2	+100[f]
FCCP (10 ug/ml)	23.7±5.0	135.0±24
Atractyloside (50 uM)	20.4±1.6	N.A.[g]
Trypsin (1 mg/ml)	116.1±3.8	17.2±1.6
Trypsin (1 mg/ml)+ Trypsin Inhibitor (1.5 mg/ml)[e]	106.9±2.1	58.7±3.1
External Trypsin (1 mg/ml)[e]	101.6±2.8	115.8±6.0

[a]Disruption of intact synaptosomes was accomplished by saponin pretreatment at 250
g/ml for 30 min at 4°C.
[b]Mitochondrial Mg-ATP hydrolysis activity was determined as the mitochondrial poison-
sensitive aspect of the ATP hydrolyzing activity apparent in the presence of Basal
Mg-ATP hydrolysis buffer system (see footnote g in Table 2 and Table 1,
respectively).
[c]Nonmitochondrial Ca+Mg ATP hydrolysis activity (see Table 1) refers only to Ca²⁺-
stimulated ATP hydrolyzing activity (described in footnote f of Table 2).
[d]percent of control refers to the percentage of activity remaining in the presence of
inhibitor compared to untreated sample. Percentages are expressed as the means ±
standard error of two separate experiments (four replicates).
[e]Trypsin inhibitor was added simultaneously with trypsin. External trypsin refers to
trypsin added prior to saponin disruption.
[f](+)denotes stimulation above untreated control value, see Fig. 2.
[g]Not available (N.A.).

drial activity. External trypsin treatment had little effect on
either system.

Influence of Agents Modifying Ionic Fluxes Across Biomembranes

The results summarized in Table 4 reveal that many of the agents
examined produced only marginal effects to be of practical use under
these experimental conditions. However, treatment with the K⁺
ionophore, valinomycin, resulted in substantial inhibition in all
preparations and in particular to nonmitochondrial Ca+Mg ATP
hydrolysis measured in the synaptic preparation. Treatment with PTZ
has been reputed to cause increased bursting activity in ganglionic
nerve cells and this phenomena has been attributed to a release of
intracellular Ca²⁺ stores (Sugaya and Onozuka, 1978). It is interesting that PTZ is particularly inhibitory to nonmitochondrial Ca+Mg
ATP hydrolysis in disrupted synaptosomes from insect brain. Mersalyl
acid is inhibitory to all ATP hydrolyzing activities but particularly
to nonmitochondrial Ca+Mg ATP hydrolysis. Tetracaine, a local

Table 4. Influence of agents modifying Na[+], K[+], or Ca[2+] flux on
Ca[2+]-stimulated ATP hydrolyzing activities[a] from microsomal
and synaptosomal fractions of the insect brain

Additions	1 mM Na-Ca ATP Hydrolysis		Nonmitochondrial Ca+Mg ATP Hydrolysis[b]	
	Microsomal	Synaptosomal	Microsomal	Synaptosomal
	Percent of control + S.E.[c]			
Na[+] flux modifiers				
veratridine (200 uM)	90.4+1.7	78.8+4.6	83.3+6.8	73.6+0.8
ouabain (.1 mM)	105.0+7.1	98.2+3.0	111.0+10.7	101.0+2.5
TTX (5 nM)	94.9+2.5	85.8+9.4	81.7+1.9	96.6+6.5
K[+] flux modifiers				
valinomycin (50 uM)	55.1+7.0	47.0+0.1	63.4+1.0	15.9+1.0
Ca[2+] flux modifiers				
A23187 (10 uM)	78.2+2.3	71.0+8.7	109.0+4.6	-
PTZ (50 mM)	91.7+1.0	81.9+7.0	90.1+1.5	0.4+8.5
Tiapamil (10 uM)	98.6+2.6	86.2+0.2	78.0+3.4	91.9+6.3
TFP (10 uM)	83.2+7.1	65.6+1.5	64.7+2.9	114.0+8.4
DSP (10 uM)	95.4+0.7	79.2+0.1	67.8+2.1	79.0+2.1
Tetracaine (10 mM)			20.4+2.5	32.6+0.6
ATPase inhibitors				
Mersalyl acid (50 uM)	41.5+2.1	53.2+9.8	0.0	32.1+1.0
Colchicine (10 uM)	94.8+0.4	84.9+2.7	77.0+2.5	95.8+7.1
Cytochalasin B (10 uM)	101.0+1.9	79.7+2.0	67.3+2.4	113.0+3.7

[a]Ca[2+]-stimulated ATP hydrolysis activity refers to the ATP hydrolyzing activity
determined as the difference between Total activity (plus Ca[2+]) and Basal activity
(minus Ca[2+]).
[b]Nonmitochondrial Ca+Mg ATP hydrolysis activity refers only to Ca[2+]-stimulated ATP
hydrolyzing activity (described in footnote f of Table 2) in the combined presence
of mitochondrial poisons (see Table 1 and footnotes c and h in Table 2).
[c]Percent of control refers to the percentage of Ca[2+]-stimulated activity remaining in
the presence of an added agent when compared to an untreated control. Percentages
are expressed as means + standard error of two separate experiments (four
replicates).

anesthetic, has been reported to cause a large decrease in ATP-
dependent Ca[2+] uptake by nonmitochondrial entities in synaptosomes
(i.e., endoplasmic reticulum, Blaustein et al., 1978a).   Tetracaine
produced a similar decrease in nonmitochondrial Ca+Mg ATP hydrolysis
as measured in both microsomal and synaptosomal tissue fractions.

Figure 1 illustrates the effects of two additional ionic flux
modifiers on ATP hydrolyzing activities. It is apparent that
monensin, a Na[+] ionophore, is increasingly inhibitory to Na-Ca ATP
hydrolysis as the Ca[2+] concentration is reduced from 50 µM to 0.5 µM.
Over the same concentration range, monensin is without effect on
nonmitochondrial Ca+Mg ATP hydrolysis. Conversely the trivalent

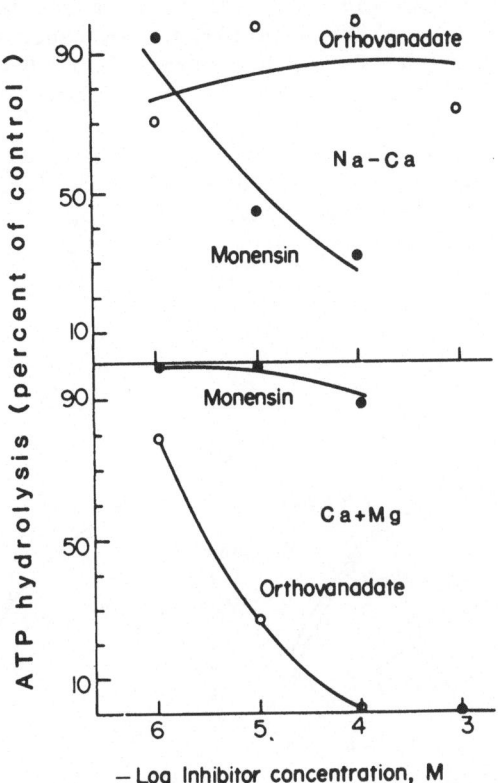

Fig. 1. Inhibition of microsomal Ca^{2+}-stimulated ATP hydrolysis by
monensin (O) and orthovanadate (o). ATP hydrolysis is
reported on a relative basis as the percentage of activity
remaining of an untreated control value taken as equal to
100% for each system examined. Only the Ca^{2+}-stimulated
portion of each ATP hydrolyzing activity was examined.
Ca+Mg ATP hydrolysis refers only to nonmitochondrial activity.

anion, orthovanadate, is particularly inhibitory to Ca+Mg ATP
hydrolysis. Gill et al. (1981) has determined that monensin is a
selective inhibitor of Na/Ca exchange in membrane vesicles from
guinea pig brain synaptosomes whereas orthovanadrate is principally
inhibitory to ATP-dependent Ca^{2+} uptake into these membrane vesicles.

A number of reports have indicated that there are other cellular
aspects besides the Ca^{2+} uptake by mitochondria which are sensitive
to ruthenium red. Watson et al. (1971) found that the calcium pump
of the red blood cell is inhibited by ruthenium red ($I_{50}=10^{-5}$M).
Rahamimoff and Alnaes (1973) concluded that the main action of
ruthenium red (10^{-5}M) on synaptic transmission is at the presynaptic
nerve terminal where it decreased the number of quanta of transmitter

liberated by the nerve impulse. Ruthenium red also increased the
frequency of minature end plate potentials. As represented in Fig.
2, ruthenium red reduces Na-Ca ATP hydrolysis in both microsomal and
synaptosomal fractions at concentrations at or above 10^{-5}M.
Ruthenium red produced a greater effect at lower Ca^{2+} concentrations
and is noticeably more inhibitory when applied to synaptic preparations. Similarly, nonmitochondrial Ca+Mg ATP hydrolysis is significantly reduced by ruthenium red concentrations of 10^{-5}M or greater
but in the synaptic fraction this activity is enhanced by low levels
of ruthenium red (i.e., 10^{-8} to 10^{-6}M).

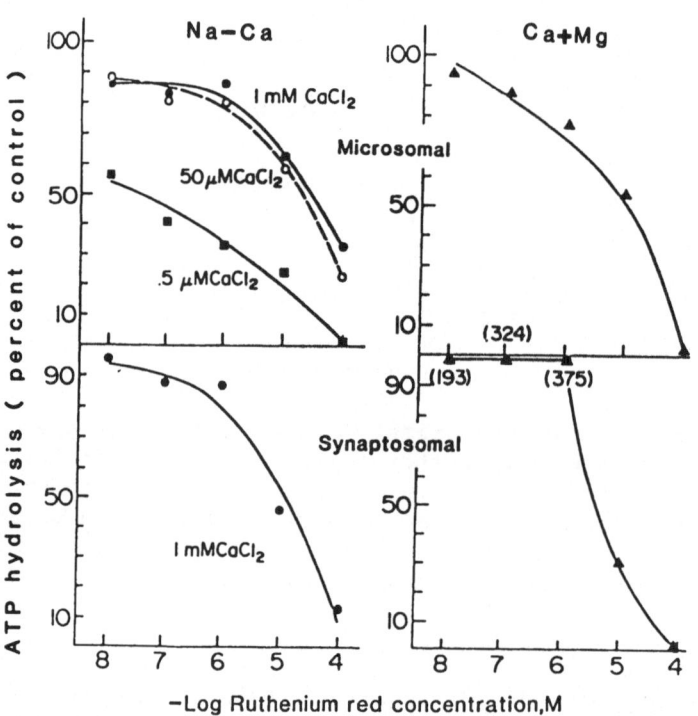

Fig. 2. Effect of ruthenium red concentration on Na-Ca ATP hydroly-
         sis at three concentrations of Ca^{2+}; 1mM (●), 50 uM (○), and
         0.5 uM(■) and on nonmitochondrial Ca+Mg ATP hydrolysis (▲).
         ATP hydrolysis is reported on a relative basis as the
         percentage of activity remaining of an untreated control
         value taken as equal to 100%. Only the Ca^{2+}-stimulated
         portion of each ATP hydrolyzing activity was examined for
         ruthenium red sensitivity in both microsomal and disrupted
         synaptosomal tissue fractions. The numbers in parentheses
         indicate the percent stimulation value of a treated sample
         above an untreated control.

Determination of Pyrethroid Action on Ca^{2+}-stimulated ATP Hydrolysis
in Insect Brain

In both microsomal and synaptosomal fractions (Fig. 3), permethrin (10^{-6}M) is consistently more inhibitory as Ca^{2+} concentrations
are reduced. This is particularly true when Ca^{2+} concentrations are
reduced to a micromolar range or below. Comparatively, permethrin is
more inhibitory when applied to synaptosomal fractions than to
microsomal fractions. However, nonmitochondrial Ca+Mg ATP
hydrolysis is reduced to a greater extent by permethrin than is Na-Ca
ATP hydrolysis regardless of whether the comparison is between microsomal or synaptosomal preparations.

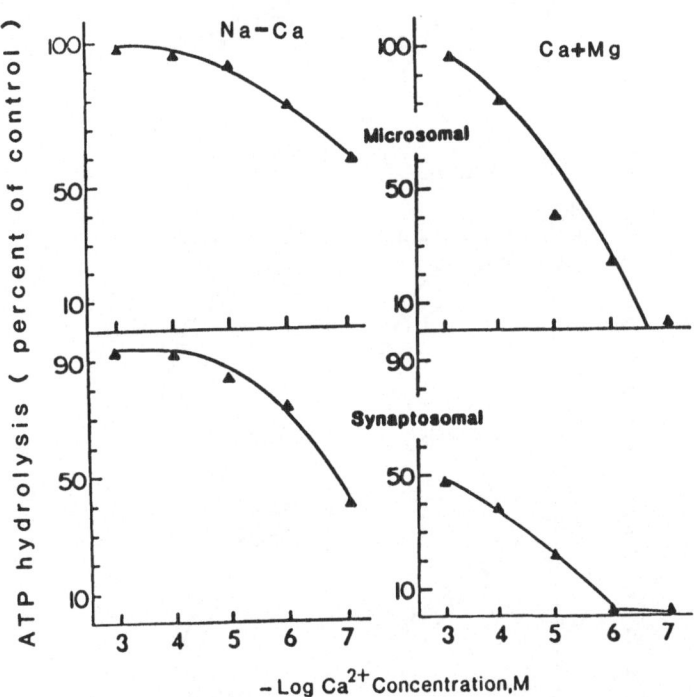

Fig. 3.   Effect of Ca^{2+} concentration on the inhibitory response of
          permethrin (10^{-6}M) on Ca^{2+}-stimulated ATP hydrolyzing
          activities. The data is expressed as the percentage of
          activity remaining in the presence of permethrin when
          compared with an untreated control value which is taken to
          be equal to 100% at each Ca^{2+} concentration examined. Na-Ca
          ATP hydrolysis was determined in microsomal and intact
          synaptosomal fractions. Nonmitochondrial Ca+Mg ATP
          hydrolysis was determined in microsomal and disrupted synap-
          tosomal factions. Only the Ca^{2+}-stimulated portion of each
          ATP hydrolyzing activity was examined for permethrin
          inhibition.

It is apparent from Fig. 4 that in addition to permethrin, allethrin and cypermethrin are also increasingly inhibitory to Na-Ca ATP hydrolysis as Ca^{2+} concentrations are reduced. Also, all these insecticides are more inhibitory when applied to synaptic preparations. However, allethrin is overall more inhibitory to Na-Ca ATP hydrolysis than either permethrin or cypermethrin and this tendency is consistent regardless of which tissue fraction is examined.

Similar experiments were performed with Ca+Mg ATP hydrolysis and as seen in Fig. 5, all three pyrethroids are more inhibitory to

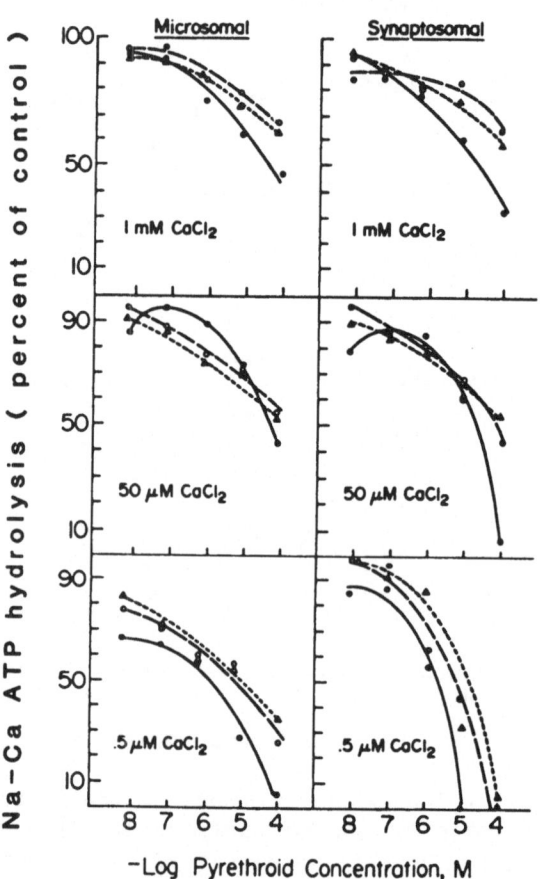

Fig. 4.   Inhibition of Na-Ca ATP hydrolysis by three pyrethroid insecticides allethrin (●), permethrin (o) and cypermethrin (Δ). ATP hydrolysis is reported on a relative basis as the percentage of activity remaining of an untreated control value taken as equal to 100% for each Ca^{2+} concentration examined. Only the Ca^{2+}-stimulated portion of each ATP hydrolyzing activity was examined for pyrethroid inhibition in microsomal and intact synaptosomal tissue fractions.

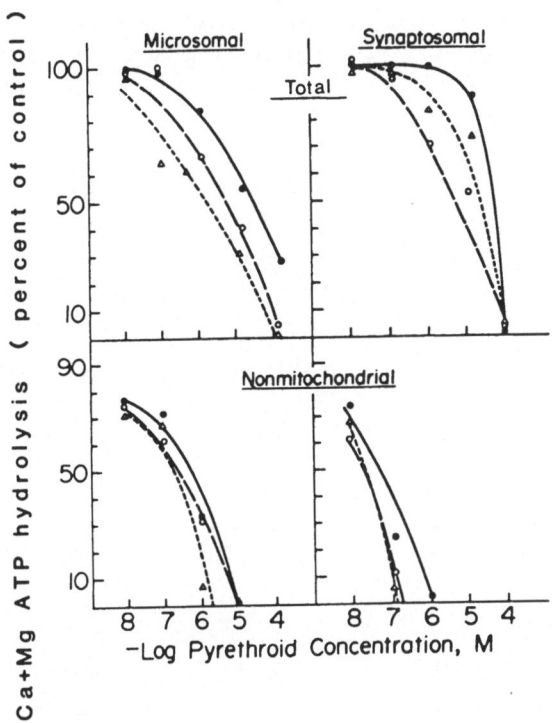

Fig. 5. Inhibition of Ca+Mg ATP hydrolysis by three pyrethroid
         insecticides; allethrin (●), permethrin (○) and cypermethrin
         (△). ATP hydrolysis is reported on a relative basis as the
         percentage of activity remaining of an untreated control
         value taken as equal to 100%. Only the Ca^{2+}-stimulated
         portion of each ATP hydrolyzing activity was examined for
         pyrethroid inhibition in microsomal and disrupted synapto-
         somal tissue fractions. Total and nonmitochondrial ATP
         hydrolyzing activities were determined exactly as previously
         outlined (refer to footnotes e and h in Table 2).

nonmitochondrial activity than to total activity. As before, the
synaptosomal preparation is the more susceptible tissue fraction. It
is also apparent that Ca+Mg ATP hydrolysis is relatively more succep-
tible to the inhibitory action of permethrin and cypermethrin than to
allethrin.

The data presented in Table 5 demonstrate that _in vivo_-
administered permethrin results in a reduction of all Ca^{2+}-stimulated
ATP hydrolytic activities when compared to control values. It is
also evident that _in vivo_-administered permethrin is overall more
inhibitory in synaptic fractions then to microsomal fractions.
Permethrin is relatively more inhibitory to Ca+Mg ATP hydrolysis than
to Na-Ca ATP hydrolysis at similar Ca^{2+} concentrations regardless of

Table 5. Effects of _in vivo_[a] administered permethrin on microsomal
and synaptosomal ATP hydrolyzing activities of the insect
brain

ATP hydrolysis buffer	Tissue fraction	Ca^{2+}-stimulated ATP hydrolyzing activity + S.E.[b] (pmole Pi/mg protein/10 min)		
Na-Ca ATP hydrolysis		Untreated	Treated	Percent of Control[c]
1 mM Na-Ca	Microsomal	3272+220	2558+32	78.2
50 uM Na-Ca	Microsomal	1968+232	1533+36	77.9
0.5 uM Na-Ca	Microsomal	179+33	92+22	51.4
1 mM Na-Ca	Synaptosomal	1683+160	1113+16	67.1
50 uM Na-Ca	Synaptosomal	1147+240	754+50	65.8
0.5 uM Na-Ca	Synaptosomal	151+19	0+0	0
Nonmitochondrial Ca+Mg ATP hydrolysis[d]				
10 uM Ca	Microsomal	322+60	114+9	35.3
10 uM Ca	Synaptosomal	141+19	25+30	17.9

[a]For _in vivo_ studies, a LD_{95} dose of permethrin was injected intraperitoneally. Once
convulsions were established (usually 1 hr after injection), the roaches were
sacrificed and brains fractionated as previously described (see Methods). Specific
activity of various ATP hydrolyzing systems was determined under standard assay
conditions (see Table 1).
[b]Specific activities of Ca^{2+}-stimulated ATP hydrolyzing activities (see footnote f in
Table 2) are expressed as the means (pmole Pi/mg protein/10 min) + standard error
(S.E.) from two separate experiments for treated samples and from eight separate
experiments for untreated control samples.
[c]Data are expressed as percentage of Ca^{2+}-stimulated ATP hydrolyzing activity
remaining in samples treated with permethrin versus untreated control.
[d]Nonmitochondrial Ca+Mg ATP hydrolysis refers only to Ca^{2+}-stimulated ATP
hydrolyzing activity (described in footnote f of Table 2) in the combined
presence of mitochondrial poisons (see footnotes c and h in Table 2).

the tissue fraction examined. These _in vivo_ results are consistent
with those obtained using _in vitro_ experimental conditions.

The relative amounts of permethrin bound to various brain fractions of insects severely poisoned with permethrin _in vivo_ and _in
vitro_ were also determined. As seen in Table 6, undisrupted synaptosomal fractions contained the most ^{14}C-permethrin, followed by microsomes and then whole brain when permethrin was administered _in vivo_.
These results correlate well with the data presented in Table 5 where
ATP hydrolyzing activities of the synaptosomal fraction is consistently more inhibited by permethrin than in the microsomal fraction.

Table 6.   Binding of ^{14}C-permethrin to insect brain fractions under
in vivo and in vitro conditions

Tissue fraction	In vivo[a]	In vitro[b]
	(ng permethrin/mg protein)[c]	
Whole brain	1.97[d]	
Microsomal	2.44	5.81
Synaptosomal	2.86	3.56

[a]Experimental conditions were exactly as described in footnote a of Table 5 except
^{14}C-permethrin was utilized.
[b]For in vitro studies, brain fractions were prepared from untreated roaches. An I_{50}
dose of ^{14}C-permethrin for nonmitochondrial Ca+Mg ATP hydrolysis was added during a
standard ATP hydrolysis assay (see Methods) using unlabeled ATP as substrate.
[c]Data are expressed as the amount of permethrin bound to the total tissue fraction
(20 insect brain equivalents) adjusted to a milligram protein level. Twenty insect
brains contained approximately 15.25 mg protein.
[d]Given the wet weight of a brain as 2.06 mg, the apparent cis-permethrin
concentration in the brain of a roach showing severe poisoning symptoms in vivo is
estimated at approximately 729 ppb.

However, a greater amount of ^{14}C-permethrin was found bound to the
microsomal and synaptosomal fractions under in vitro conditions than
that which had been determined previously under in vivo conditions.
This is apparent eventhough a lower amount of bound permethrin gave a
relatively greater amount of inhibition in vivo (see nonmitochondrial
Ca+Mg ATP hydrolysis data, Table 5).   This descrepancy may be due in
part to a greater amount of nonspecific binding in the more disrupted
in vitro prepatations. Nonetheless, the amount of ^{14}C-permethrin
bound in vitro is of the same magnitude as that determined under in
vivo conditions indicating that the level of permethrin in the nervous system at the time of severe permethrin poisoning is similar to
the amount required to cause significant inhibition (e.g., I_{50}) of
Ca²⁺-stimulated ATP hydrolysis activites in vitro.

DISCUSSION

    The above studies with inhibitors and agents affecting ion
transport have established that, first, mitochondrial ATP hydrolysis
could be segregated from other Ca²⁺-stimulated ATP hydrolytic
activities in the insect brain by the use of a variety of mitochondrial poisons including ruthenium red and FCCP. Second, nonmitochondrial Ca+Mg ATP hydrolysis may be recognized by its sensitivity to
trypsin digestion, orthovanadate, valinomycin and PTZ. However, it
must be noted that the last two agents work only in synaptosomes
which are expected to have retained their original structural
integrity. The selective inhibition of nonmitochondrial Ca+Mg ATP

hydrolysis by orthovanadate correlates well with its action on ATP-dependent Ca^{2+} transport in synaptolemma vesicles (Gill et al., 1981) and in squid giants axons (i.e., uncoupled ATP-dependnt Ca^{2+} transport, DiPolo, 1978; DiPolo et al., 1979). Third, Na-Ca ATP hydrolysis may be recognized by its sensitivity to monensin which is known to dissipate sodium gradients across the membrane (Gill et al., 1981). It is well documented that by removing the sodium gradient monensin also eliminates the driving force necessary for Na/Ca exchange (Mullins, 1981).

While Ca+Mg ATP hydrolysis requires the presence of K^+, Mg^{2+} and at least micromolar concentration of Ca^{2+}, there is a portion of the overall Ca^{2+}-stimulated ATP hydrolytic activity in the cockroach nervous system that is selectively enhanced by $Na^+(Li^+)$, Ca^{2+} and K^+ (designated as Na-Ca ATP hydrolysis in this paper). This Ca^{2+}-stimulated ATP hydrolyzing activity is very similar to one found in the lobster nerve (Ghiasuddin and Matsumura, 1979) and in optic nerve of squid, (Matsumura and Clark, 1980, 1982). As for the probable function of Na-Ca ATP hydrolysis, work from this laboratory using squid axons (Matsumura and Clark, 1982; Clark et al., 1985) indicates that it could be coupled to Na/Ca exchange. Although Na/Ca exchange is absolutely dependent on a transmembrane sodium gradient, ATP appears to increase the affinity of the exchanger to internal Ca^{2+} especially at low calcium concentrations (i.e., normal physiological condition, Blaustein and Santiago, 1977; Matsumura and Clark, 1982). This ATP-promoted aspect of Na/Ca exchange is realized only with ATP or other hydrolyzable ATP analogs (DiPolo, 1977). Also, the selective inhibition of Na-Ca ATP hydrolysis by monensin is identical to that observed on Na/Ca exchange by Gill et al. (1981). Indeed, recent experiments (Caroni and Carafoli, 1983) have shown a Na/Ca exchanger in heart sarcolemma to be activated by a treatment of ATP and Ca^{2+} and deactivated by treatment with an exogenously added phosphorylase phosphatase. Activation with adenosine 5'[γ-thio] triphosphate showed that the activating process is carried out by a protein kinase reaction and the deactivating process by a phosphoprotein phosphatase reaction. More interesting is the fact that both processes are modulated by Ca-calmodulin interactions. As shown in Table 4., TFP (10 μM), a phenothiazine antipsychotic that binds calmodulin specifically (Weiss et al. 1980), inhibits some 20-40% of the total Na-Ca ATP hydrolysis. With this in mind, Na-Ca ATP hydrolysis may be better defined as Na-Ca protein kinase-phosphatase activity.

As previously reported in squid (Clark and Matsumura, 1982) and now verified in the American cockroach, pyrethroid insecticides are inhibitory to Ca^{2+}-stimulated ATP hydrolytic activities. In both organisms, allethrin, a closely related synthetic analog of pyrethrin, is more inhibitory to Na-Ca ATP hydrolysis whereas highly modified pyrethroids such as cypermethrin (e.g., α-cyano containing) are more inhibitory to Ca+Mg ATP hydrolysis. Overall, Ca+Mg ATP hydrolysis appears more sensitive to pyrethroid action than Na-Ca ATP

hydrolysis although both are inhibited.   In the cockroach, the synaptic fraction is more inhibited by pryethroid action than the microsomal preparation with nonmitochondrial Ca+Mg ATP hydrolysis of
disrupted synaptosomes being the most sensitive ATP hydrolyzing
activity examined.

This may be particularly relevent in that calcium regulation in
the presynaptic nerve terminal has been determined to control the
rate of neurotransmitter release (Llinas et al., 1982).   Nonmitochondrial Ca+Mg ATP hydrolysis in the insect synaptosome has been
shown to possess many of the same characteristics determined for ATP-
dependent Ca^{2+} transport by endoplasmic reticulum and synaptic
plasma membrane in similarly distrpted synaptosomes (Blaustein et
al., 1978a; McGraw et al, 1980, Gill et al., 1981, 1984). Thus,
inhibition of ATP hydrolyzing activity in this region by pyrethroid
insecticides could lead to increased levels of ionized Ca^{2+} resulting
in the disruption of a number of Ca^{2+}-regulated processes including
increased neurotransmitter release. Indeed, it has been suggested
that the calcium pump functions primarily in the maintenance of low
resting levels of cytosolic Ca^{2+} rather than in the removal of larger
levels of Ca^{2+} which result during bioelectrical activity, a function
much more suited for the Na/Ca exchanger with its higher capacity and
slightly reduced affinity for Ca^{2+}. Therefore, disruption of the
calcium regulatory function of either of the two entities by
pyrethroid insecticides could result in the loss of calcium homeostasis in the nerve cell.

These in vitro findings have been substantiated by in vivo
experiments where permethrin was administered to live roaches. All
Ca^{2+}-stimulated ATP hydrolyzing activities which were examined were
reduced by this treatment, especially nonmitochondrial Ca+Mg ATP
hydrolysis in synaptosomes. This is the first reported incidence of
in vivo inhibition of Ca^{2+}-stimulated ATP hydrolyzing activity by
pyrethroid insecticides in organisms which clearly showed symptoms of
pyrethroid toxicity. More importantly, this in vivo inhibition
occurred in the presence of a similar amount of insecticide which was
seen to cause a substantial amount of inhibition in vitro.

As for the relationship of these calcium-dependent systems to
pyrethroid-induced nerve symptoms, Gammon (1980) has shown electrophysiologically that permethrin-resistant insects (Spodoptera
littaralis), which under normal Ca^{2+} conditions are relatively recal-
citrant to permethrin, are caused to mimic the electrical nerve
patterns of poisoning seen in permethrin-sesceptible insects by the
reduction of Ca^{2+} in the saline buffer. Similarly, there have been a
number of reports in which the effects of DDT have been altered by
calcium manipulation (Matsumura and Narahashi, 1971; Batterton et
al., 1972; Matsumura and Clark, 1985). Therefore, there is a strong
possibility that at least some portion of the symptoms caused by
pyrethroid poisoning can be explained by their biochemical affects on

Ca^{2+}-stimulated ATP hydrolyzing activities as observed in this work. This is especially true in that they appear to occur via a coupled phosphorylation – dephosphorylation event in nerve membranes which appears highly sensitive to Ca^{2+} concentration. (see Fig. 10, Matsumura and Clark, 1980).

ACKNOWLEDGEMENT

Supported by the Massachusetts Agricultural Experiment Station, University of Massachusetts, Amherst, Michigan Agricultural Experiment Station, Michigan State University, and by research grant ESO1963 from the National Institute of Environmental Health Sciences, Research Triangle Park, North Carolina.

REFERENCES

Abood, L.G., and Gerard, R.W., 1954, Enzyme distribution in isolated particulates of rat peripheral nerve, J. Cell. Comp. Physiol., 43:379-392.

Aldridge, W.N., 1980, Pyrethroid Insecticides: Chemistry and Action. Table Ronde Roussel-Uclaf No. 37.

Batterton, J.C., Boush, G.M., and Matsumura, F., 1972, DDT: inhibition of sodium chloride tolerance by the Blue-green alga, Anacystics nidulans, Science, 176:1141-1143.

Blaustein, M.P., Ratzlaff, R.W., Kendrick, N.C., and Schweitzer, E.S., 1978a, Calcium buffering in presynaptic nerve terminals; I. evidence for involvement of a nonmitochondrial ATP-dependent sequestration mechanism, J. Gen. Physiol., 72:15-41.

Blaustein, M.P., Ratzlaff, R.W., and Kendrick, N.K., 1978b, The regulation of intracellular calcium in presynaptic nerve terminals, in, "Calcium Transport and Cell Function", A. Scarpa and E. Carafoli, eds., Ann. New York Acad. Sci., 307:195-212.

Blaustein, M.P., and Santiago, E.M., 1977, Effects of internal and external cations and ATP on sodium-calcium and calcium-calcium exchange in squid axons, Biophy. J., 20:79-111.

Carafoli, E., and Crompton, M., 1978, The regulation of intracellular calcium by mitochondria, in, "Calcium Transport and Cell function", A. Scarpa and E. Carafoli, Eds., Ann. New York Acad. Sci.

Caroni, P., and Carafoli, E., 1983, The regulation of $Na^{+}-Ca^{2+}$ exchanger of heart sarcolemma, Eur. J. Biochem., 132:451-460.

Chadwick, P.R., 1979, The activity of some pyrethroids against Periplaneta americana and Blattella germanica, Pestic. Sci., 10:32-38.

Clark, J.M., and, Matsumura, F., 1982, Two different types of inhibitory effects of pyrethroids on nerve Ca- and Ca+Mg ATPase in the squid, Loligo pealei, Pestic. Biochem Physiol., 18:180-190.

Clark, J. Marshall, Jones, E.L., and Matsumura, F., 1985, Is Na-Ca stimulated ATP hydrolysis found in squid retinal nerve identical to the ATP promoted aspect of Na/Ca exchange?, Biochim. Biophy. Acta, (in press).

Desaiah, D., Cutkomp, L.K., and, Loch, R.B., 1974, A comparison of DDT and its biodegradable analogues tested on ATPase enzymes in cockroach, Pestic. Biochem. Physiol., 4:232-238.

DiPolo, R., 1977, Characteristics of the ATP-dependent calcium efflux in dialyzed squid giant axons, J. Gen. Physiol., 69:795-813.

DiPolo, R., 1978, Ca pump driven by ATP in squid axons, Nature, 274:390-392.

DiPolo, R., Rojas, H.R., and, Beange, L., 1979, Vanadate inhibits uncoupled Ca efflux but not Na-Ca exchange in squid axons, Nature, 281: 228-229.

Doherty, J.D., and Matsumura, F., 1975, DDT effects on certain ATP-related systems in the peripheral nervous system of the lobster, Homarus americanus, Pestic. Biochem. Physiol., 5:242-252.

Frankenhaeuser, B., and Hodgkin, A.L., 1957, The action of calcium on the electrical properties of squid axons, J. Physiol., 137:218-244.

Gammon, D.W., 1980, Pyrethroid resistance in a strain of Spodoptera littoralis is correlated with decreased sensitivity of the CNS in vitro, Pestic. Biochem. Physiol., 13:53-62.

Gammon, D.W., Brown, M.A., and Casida, J.E., 1981, Two classes of pyrethroid action in the cockroach, Pestic. Biochem. Physiol., 15:181.

Ghiasuddin, S.M., and Matsumura, F., 1979, DDT inhibition of Ca-ATPase of the peripheral nerves of the American lobster, Pestic. Biochem. Physiol., 10:151-161.

Gill, D.L., Grollman, E.F., and Kohn, L.D., 1981, Calcium transport mechanisms in membrane vesicles from guinea pig brain synaptosomes, J. Biol. Chem., 256:184-192.

Gill, D.L., Chueh, Seau-Huei, and Whitlow, C.L., 1984, Functional importance of the synaptic plasma membrane calcium pump and sodium-calcium exchanger, J. Biol. Chem., 259(17):10807-10813.

Huddard, H., Greenwood, M., and Williams, A.J., 1974, The effect of some organophosphorous and organochlorine compounds on calcium uptake by sarcoplasmic reticulum isolated from insect and crustacean skeletal muscle, J. Comp. Physiol., 93:139-150.

Janicki, R.H., and Kinter, W.B., 1971, DDT inhibits Na⁺, K⁺, Mg²⁺-ATPase in intestinal mucosae and gills of marine teleots, Nature 233:(39), 148.

Lenninger, A.L., 1977, Oxidative phosphorylation and mitochondrial structure, in, "Biochemistry" (2nd Ed.). New York, Worth Publishers, Inc., 530.

Llinas, R.R, 1982, Calcium in synaptic transmitter, Sci. Amer., 287(4),56-194.

Lowry, O.H., Rosenbrough, N.J., Farr, A.L., and Randall, R.J., 1951, Protein measurement with folin phenol reagent, J. Biol. Chem.,

193:265-275.

Matsumura, F., 1977, Ion-sensitive ATP-phosphorylation processes in the axonic membrane of squid retinal nerve, Comp. Biochem. Physiol., 58:13-20.

Matsumura, F., and Clark, J. Marshall, 1980, ATPase in the axon-rich membrane preparation from the retinal nerve of the squid, Loligo pealei, Comp. Biochem. Physol., 66B:23-32.

Matsumura, F., and Clark, J.M., 1982, ATP-utilizing systems in the squid axons: A review on the biochemical aspects of ion-transport, Progress in Neurobiology., 18(4):231-256.

Matsumura, F., and Clark, J. Marshall, 1985, Investigations on the suitability of using nerve membrane fragments incorporated into artificial liposomes as a method for the study of pesticidal action on sodium channel activity, NeuroToxicology, 6(2):271-288.

Matsumuara, F., and Hayashi, M., 1969, Comparative mechanisms of insecticide binding with nerve components of insects and mammals, Residue Reviews, 25:265-273.

Matsumura, F., and Narahashi, T., 1971, ATPase inhibition and electrophysiological changes caused by DDT and related neuroactive agents in lobster nerve, Biochem. Pharmacol., 20:825-837.

Matsumura, R., and Patil, K.C., 1969, Adenosine triphosphotase sensitive to DDT in synapse of rat brain, Science, 166:121-122.

McGraw, C.F., Somlyo, A.V., and Blaustein, M.P., 1980, Localization of calcium in presynaptic nerve terminals, J. Cell Biol., 85:228-241.

Miller, D.S., Seymour, A., Shoemaker, D., Winesor, M.H., Peakall, D.B., and Kinter, W.B., 1976, Possible enzymatic basis of DDT-induced eggshell thinning in the white Peking duck, Anas platyrhynchos. Bull Mt. Desert Isl. Biol. Lab., 14:73-76,

Mullins, L.J., 1981, Ca entry upon depolarization of nerve, J. Physiol., Paris, 77:1139-1144.

Narahashi, T., 1979, Nerve membrane ionic channels as target site of insecticidem in, "Neurotoxicology of Insecticides and Phero-mones", T. Narahashi, ed., Plenum Press, New York.

Nishimura, K., and Narahashi, T., 1978, Structure-activity relation-ships of pyrethroids based on direct action on nerve, Pestic. Biochem. Physiol., 8:53.

Rahamimoff, R., and Alnaes, E., 1973, Inhibitory action of ruthenium red on neuromuscular transmission, Proc. Natl. Acad. Sci., USA 70(12):3613-3616.

Rahamimoff, R., Eurlkar, S.D., Alnaes, E., Meiri, H., Rotshenker, S., and Rahamimoff, H., 1976, Modulation of transmitter release by calcium ions and nerve impulses, in, "The Synapse", XL Cold Spring Harbor Symposia on Quantative Biology, 107-116.

Reinlib, L., Caroni, P., and Carafoli, E., 1981, Studies on heart sarcolemma: vesicles of opposite orientation and effect of ATP on the Na^+/Ca^{2+} exchanger, FEBS Letters, 126(1):74-76.

Schneider, R.P., 1975, Mechanism of inhibition of rat brain (Na+K)-adenosine triphosphatase by 2,2-bis (p-chlorophenyl)-1,1,1-

trichloroethane (DDT), Biochem. Pharmacol., 24:939-346.

Skou, J.C., 1957, The influence of some cations on an adenosine triphosphatase from peripheral nerves, Biochem. Biophys. Acta, 23:394-401.

Sugaya, E., and Onozuka, M., 1978, Intracellular calcium: Its release from granules during bursting activity in snail neurons, Science 202:1195-1197.

Tasaki, I., 1974, Energy transducation in the nerve membrane and studies of excitation processes with extrinsic fluorescent probes, Ann. N.Y. Acad. Sci., 227:247.

Telford, J.N., 1968, Mechanism of dieldrin-resistance in the German cockroach. Ph.D. thesis Univ. of Wisconsin, Madison, Wisconsin. van Breemen, C., and de Weer, P., 1970, Lanthanum inhibition of ⁴⁵Ca efflux from squid giant axon, Nature, London, 226:760-761.

Vijverberg, H.P.M., and van den Bercken, J., 1979, Frequency-dependent effects of the pyrethroid insecticide decamethrin in frog myelinated nerve fibres, Eur. J. Pharmac., 58:501-504.

Watson, E.L., Vincenzi, F.F., and Davis, P.W., 1971, Ca²⁺ activated membrane ATPase: Selective inhibition by ruthenium red, Biochem. Biophys. Acta, 249:606-610.

Weiss, B., Prozialeck, W., Cimino, M., Barnette, M.S., and Wallace, T.L., 1980, Pharmacological regulation of calmodulin, Ann. N.Y. Acad. Sci., 77(8923):256-319.

Yamaguchi, I., Matsumura, F., and Kadous, A.A., 1979, Inhibition of synaptic ATPases by heptachlorepoxide in rat brain, Pestic. Biochem. and Physiol., 11:285-293.

THE NICOTINIC ACETYLCHOLINE RECEPTOR: MOLECULAR ASPECTS AND INTERACTIONS WITH INSECTICIDES

Mohyee E. Eldefrawi, Shebl M. Sherby
and Amira T. Eldefrawi

Department of Pharmacology and
Experimental Therapeutics,
University of Maryland School of Medicine
Baltimore, Maryland 21201

INTRODUCTION

Acetylcholine (ACh) receptors in vertebrates are found in
muscles, glands, brain and spinal cord, whereas in insects they are
found mainly in brain and ganglia. There are two major classes of
ACh receptors. Nicotinic ACh (nACh) receptors, which are activated
by ACh and nicotine and inhibited by d-tubocurarine (dTC) and α -
bungarotoxin (α-BGT), are found in vertebrates mainly at the neuromuscular junction of skeletal muscles and in autonomic ganglia. The
second class is muscarinic ACh (mACh) receptors, which are activated
by ACh and muscarine and inhibited by atropine and quinuclidinyl
benzilate (QNB), and are found mainly in smooth and cardiac muscles,
glands and brain of vertebrates. There are subtypes for each of
these classes. ACh receptors mediate many cellular responses, and
their function is modified by numerous drugs, including therapeutics.

Both classes of ACh receptors are present in the central nervous
system (CNS) of insects since nicotinic and muscarinic drugs have
been found to inhibit the central excitatory action of ACh (Shankland
et al., 1971; Harrow and Sattelle, 1983). In addition, nACh and mACh
receptors have been identified biochemically in insect neural tissue
by means of their specific binding of radioactive α-BGT and QNB,
respectively (Eldefrawi and Eldefrawi, 1980, 1983). nACh receptors
are recognized as the primary target for the insecticidal action of
nicotine and other nicotinoids (Yamamoto, 1965; Eldefrawi et al.,
1970). They may also be primary targets for cartap and its parent
natural neurotoxin, nereistoxin (Sakai, 1966, 1967; Bettini, 1973;
Eldefrawi et al., 1980a). In addition, nACh and mACh receptors are

213

involved indirectly in the toxic action of carbamate and organophosphate insecticides, since it is receptor desensitization which
results from accumulation of the neurotransmitter ACh after inhibition of ACh-esterase that ultimately blocks cholinergic transmission.

This presentation will concentrate on the interaction of
insecticides with nACh receptors of vertebrates and insects, using
membrane preparations and new technologies (e.g., receptor binding
and tracer ion flux assays) that detect changes in receptor numbers,
drug affinities and function which result from their binding of
insecticide.

REVIEW OF RELATED WORK

The nACh receptor is an integral membrane glycoprotein, which is
a regulatory ion transport system that is activated when ACh binds to
its specific "receptor sites." It is found in micromolar concentrations (1 nmole/g tissue) in electric organs of the electric ray
Torpedo sp. (O'Brien et al., 1972), and the pure receptor protein was
isolated by affinity chromatography from detergent extracts
(Eldefrawi and Eldefrawi, 1973; also see reviews by Heidmann and
Changeux, 1978; Karlin, 1980; Lindstrom and Dau, 1980). The receptor
protein of electric organs is made up of five subunits (2α, 1β, 1γ,
1δ) (Raftery et al., 1980), each of which is exposed extracellularly
and intracellularly (Hartig and Raftery, 1977; Strader and Raftery,
1980). The smallest subunit (α, MW 40 K) carries the recognition
(receptor) site for ACh agonists and competitive antagonists. The
two α subunits and the other three, β, γ and δ, contribute to the
structure of the ion channel that traverses the membrane, since
noncompetitive inhibitors of its function bind to 'channel sites' on
one or more of the five subunits (Oswald and Changeux, 1981). These
drugs do not block access of ACh to its receptor site, and their
effects are time and voltage dependent. Examples are the local
anesthetic procaine (Katz and Miledi, 1975), the antiviral drug
amantadine (Alburquerque et al., 1978), the general anesthetic
phencyclidine (PCP) (Alburquerque et al., 1980) and the saturated
spiropiperidine analog of histrionicotoxin which is isolated from the
skin of the Colombian frog Dendrobates histrionicus (Eldefrawi et
al., 1977), perhydrohistrionicotoxin (H_{12}HTX). It inhibits receptor-
regulated $^{22}Na^+$ transport and ACh sensitivity at the neuromuscular
junction without affecting binding of $[^3H]$ACh to Torpedo ACh
receptors (Eldefrawi et al., 1980b) or $[^3H]$ α-BGT to skeletal muscle
(Dolly et al., 1976). Histrionicotoxin and amantadine have been
shown to cause similar voltage-dependent inhibition of nACh receptors
in the CNS of insects (Sattelle and David, 1983; Artola et al.,
1984).

Membrane-bound nACh receptors of the electric organs of Torpedo
and Electrophorus and mammalian muscle were identified by their

reversible binding of agonists (e.g., [^{3}H]muscarone, [^{3}H]nicotine) and antagonists (e.g., [^{14}C]dTC) (Eldefrawi et al., 1971a,b) as well as the quasi-irreversible binding of [^{125}I] α-BGT (Bradley et al., 1976).

On the basis of their specific binding of [^{125}I] α-BGT, putative nACh receptors have been identified in house fly brain Musca domestica (Eldefrawi and Eldefrawi, 1980; Jones et al., 1981), in the antennal lobes of the moth Manduca sexta (Hildebrand et al., 1979), in the nerve cord of the American cockroach (Periplaneta americana) (Sattelle et al., 1980) and in Drosophila melanogaster (Schmidt-Nielsen et al., 1977; Dudai, 1978). These nACh receptors of insect nerve tissues have similar drug specificities to the vertebrate neuromuscular nACh receptor, except that the insect receptors have much lower affinities for decamethonium and other bis-quaternary drugs as well as for ACh and other choline esters (Eldefrawi and Eldefrawi, 1980; Jones et al., 1981; Schmidt-Nielsen et al., 1977).

In addition, putative mACh receptors have been identified in Drosophila (Haim et al., 1979) and house fly brain (Jones and Sumikawa, 1981; Shaker and Eldefrawi, 1981). Again there are some differences between these insect receptors and mammalian brain muscarinic receptors. The insect receptors have higher affinity for dTC, and the Hill coefficients for their agonist and antagonist binding are <1, while in mammals the Hill coefficient is 1 for antagonists and <1 for agonists (Birdsall and Hulme, 1976). The lower than 1 value has recently been found to be due to subtypes of muscarinic receptors and suggests that binding of [^{3}H]QNB may distinguish two muscarinic receptors in insect brain. Most interesting is the high ratio of nicotinic to muscarinic receptors in insect brain and of muscarinic to nicotinic receptors in mammalian brain (Dudai and Ben-Barak, 1977; Shaker and Eldefrawi, 1981).

Recently, we introduced the binding of [^{3}H]H$_{12}$-HTX and [^{3}H]PCP as allosteric probes whose binding is dependent on the conformation of the receptor. These ligands do not bind to the receptor site, but rather to sites on the ionic channel moiety, and the kinetics of their binding is extremely sensitive to conformational changes (Eldefrawi et al., 1980b,c). These probes also bind to house fly head extracts, but their binding is mostly to non-ACh-receptor sites (Eldefrawi et al., 1982a), even though their synaptic effects on insect neurons is similar to theirs on vertebrate nicotinic synapses (Artola et al., 1984).

Several insecticides have been shown to interact with the nACh receptor. These include organophosphate and carbamate anticholinesterases, which inhibit binding of [^{3}H]ACh to the Torpedo nicotinic receptor, but with low affinity (Eldefrawi et al., 1971; 1982a). Also, nereistoxin, which was isolated from the marine annelid Lumbriconereis heteropoda, and is toxic to insects, particularly leaf-chewing species (Sakai, 1964), acted as a partial agonist and inhibited [^{3}H]ACh binding to Torpedo nACh receptors and endplate currents in frog and rat skeletal muscles (Eldefrawi et al., 1980a). Thus, the insecticide cartap, which is converted to nereistoxin in vivo and blocks synaptic transmission in the cockroach ganglia (Bettini et al., 1973), may be acting by inhibiting central nACh receptors.

Binding of receptors ligands (e.g., [^{125}I]ACh) to nACH receptor under equilibrium conditions can identify only interactions that occur at the receptor site by agonists and competitive antagonists. Since a desensitized receptor has a higher affinity for agonists than the active receptor, the increased potency of an agonist in inhibiting [^{125}I] α-BGT is detectable by kinetic measurements of [^{125}I] α-BGT binding. It has been used successfully in studying the effect of agonists (Weiland et al., 1976) and channel drugs such as local anesthetics (Blanchard et al., 1979) or alcohols (Young and Sigman, 1981) in inducing receptor desensitization. Kinetic measurements of [^{3}H]H$_{12}$-HTX or [^{3}H]PCP binding are also more informative than equilibrium measurements, since their binding to resting nACh receptors is very slow (reaching equilibrium after 2 hours at 21°C), while binding to activated receptors (i.e., in presence of agonist) reaches equilibrium within a few minutes (Aronstam et al., 1981). We have used this assay successfully to detect interactions of quaternary ammonium anticholinesterases and pyrethroids with the 'receptor' as well as 'channel' sites of the nACh receptors (Bakry et al., 1982; Abbassy et al., 1982, 1983a,b).

Receptor-regulated tracer ion flux is used as a biochemical correlate of in vivo receptor function. Introduced by Changeux and coworkers (Kasai and Changeux, 1971a,b), the assay was modified subsequently (Miller et al., 1978; Hess et al., 1975; Epstein and Racker, 1978). Tracer ion measurements are very informative because they detect not only the interactions but its effect on the function of the receptor as well.

EXPERIMENTAL

Tissue Preparations

Torpedo electic organ. The electric organ of Torpedo is an excellent source for nACh receptors of vertebrates, because of their high concentration and the almost identical drug specificity to the nACh receptors of mammalian motor endplate. Also, frozen electric organs store well at -90°C for months.

Frozen electric organs are minced and homogenized in an equal volume buffered saline (154 mM NaCl, 5 mM Na$_2$HPO$_4$, pH 7.4, 1 mM EDTA); 0.1 mM PMSF and/or DFP may be added to the homogenization

buffer depending on the objectives of the experiment. The homogenate
is centrifuged at low speed (3000 rpm) for 10 min at $4^{\circ}C$, the pellets
rehomogenized in same buffer, recentrifuged, the two superntant fractions pooled and recentrifuged at 17,000 rpm for 60 min. The pellets
are suspended in buffer containing 0.2% NaN_3 so that the tissue
preparation contains 1 mg protein/ml, as determined by the method of
Lowry et al. (1951) using bovine serum albumin as a standard. This
membrane preparation is stored over ice in the refrigerator and is
stable for a week for binding studies.

For tracer ion measurements, the electric organ membrane is
homogenized in a high K^+ low Na^+ buffer (95 mM KCl, 5 mM NaCl and 10
mM Tris-HCl), pH 7.0, but the final membrane pellets are suspended in
a high Na^+ low K^+ buffer (95 mM NaCl, 5 mM KCl, 1.8 mM $CaCl_2$, 10 mM
Tris-HCl, pH 7.0) and the volume adjusted so that 1 ml contains 10 mg
protein. This results in membrane microsacs that have high Na^+ on
the outside and high K^+ inside as occurs in vivo.

Insect brain nACh receptor. House fly and honey bee heads are
harvested rather easily by subjecting dry-ice-frozen insects to mild
shaking over appropriate mesh sieves. For house fly heads a 7-mesh
sieve and for the larger honey bee head a 5-mesh sieve are used. The
heads are homogenized in ice-cold 0.32 M sucrose in 50 mM Na_2HPO_4,
pH 7.4 (20 w/v), the homogenate filtered through four layers of
cheesecloth to remove most of the integumental debris and the
filtrate centrifuged at 3000 x g for 10 min. The supernatant fraction is centrifuged at 30,000 rpm for 60 min, and the pellets
suspended in 1 mM phosphate buffer, pH 7.4, so that each ml contains
membranes from 0.25 g of heads. This tissue preparation may be
lyophilized, the dry powder kept for weeks at $-12^{\circ}C$ and reconstituted
in H_2O prior to assay for nACh receptor. It is important to note
that proteases are much more prevalent in the insect preparation than
the Torpedo one; thus it is necessary to have in the preparation
buffer antiproteases such as PMSF, pepstatin, trypsin inhibitor, and
aprotinin.

Binding Assays

Binding of [^{125}I] α-BGT. [^{125}I] α-BGT (sp. act. >100
Ci/mmole, New England Nuclear) binding to nACh receptors is measured
by the method of Kohanski et al. (1977), where glass disposable
Pasteur pipettes are filled at room temperature with carboxyl methyl
cellulose (Whatman-52 microgranular), 1.25 ml of settled bed volume
each, preswollen and equilibrated in phosphate buffer (1.0 mM
Na_2HPO_4, 0.01% Triton X-100 (v/v), 0.03% NaN_3, pH 7.2). Tissue
aliquots are incubated with [^{125}I] α-BGT, in the presence or absence
of drugs or insecticides, for 60 min for equilibrium measurements of
ligands that bind to the receptor site and displace α-BGT. However,
for ligands that bind to allosteric channel sites, the initial rate
of [^{125}I) α-BGT binding is measured by determining amount bound taken

during a 10-min incubation. The incubation mixture is transferred to the mini-column, is washed with 1.0 ml buffer and the eluate is collected and counted in an auto gamma counter. Unreacted $[^{125}I]$ α-BGT is adsorbed on the column, while membrane-bound $[^{125}I]$ α-BGT is eluted. Specific binding is the difference between total binding and that obtained in presence of 10 μM of the α-toxin of the venom of the cobra (Naja naja siamensis).

Binding of $[^3H]ACh.$ nACh receptors of Torpedo membranes are also identified by the reversible binding of $[^3H]ACh$ measured by equilibrium dialysis (Eldefrawi et al., 1971b). Tissue samples (250 μl), in dialysis sacs tied at both ends, are placed in 10 ml of buffered Ringer containing $[^3H]ACh$ (0.1 to 1.0 μM). Equilibration is completed by shaking at 21°C for 4 h. Drugs or insecticides to be tested are placed in the dialysis bath with $[^3H]ACh$. Three 50-μl aliquots are taken from the membrane sample and three from the dialysis bath and their radioactivity determined by liquid scintillation counting. Specific binding is totally inhibited with 10 μM Naja - α toxin.

Binding of $[^3H]$ H$_{12}$-HTX. $[^3H]H_{12}$-HTX (sp. act. 21 Ci/mmol) is added to the membrane prepration and specific binding is measured by a filter assay as described (aronstam et al., 1981). Binding to inactive nACh receptors is measured by incubating Torpedo membranes with 10 μM Naja α-toxin for 60 min which causes inhibition of the receptor sites for the duration of the experiment and guarantees that the ionic channel is closed without affecting the kinetics of $[^3H]H_{12}$-HTX binding. Membranes (50 μl) are incubated for 120 min with 950 ml of buffer (50 mM Tris buffer or phosphate Ringer) containing the test compound. The mixture is then filtered over Whatman GF/B filter (pretreated with 0.01% polylysine to reduce nonspecific binding to filters). The filter is washed with 10 ml cold buffer and placed in a minivial with 5 ml scintillation solution and the radioactivity counted in a liquid scintillation spectrometer after 8 h. Specific binding is the difference between binding of $[^3H]H_{12}$-HTX in absence and presence of 1 mM amantadine, which blocks totally specific binding to the ionic channel (Albuquerque et al., 1978).

Binding to activated nACh receptors is determined as described above except that 100 μM carbamylcholine is present during incubation with $[^3H]H_{12}$-HTX and filtration is after 30 sec. Other receptor agonists at different concentrations may be used if the purpose of the experiment so requires. $[^3H]PCP$, which is available commercially, may be substituted for $[^3H]H_{12}$-HTX and will yield similar results.

Tracer Ion Uptake Measurements

A biochemical correlate of the measurement of nACh receptor-mediated conductance changes is the measurement of its regulated ionic fluxes. Torpedo membranes (2 mg protein) are added to a test tube containing 1μ Ci ^{22}NaCl or ^{45}CaCl in 100 μl of 95 mM NaCl, 5 mM KCl, 1.8 mM CaCl$_2$, pH 7.0, containing 100 μM carbamylcholine, mixed and incubated for 60 or 15 sec, respectively. Uptake of the radioactive cation is measured by the method of Epstein and Racker (1978), where separation of free from sequestered or bound ^{22}Na or ^{45}Ca is achieved by passage through a minicolumn packed with Dowex 50W-X8 resin (100-200 mesh) equilibrated in Tris-HCl, pH 7.0. The tissue is washed through the column with 0.8 ml buffer and the eluate collected and counted in an auto gamma counter in the case of ^{22}Na or in a liquid scintillation counter for ^{45}Ca.

Drugs or insecticides to be tested for their effects on the receptor-regulated cation uptake are added to the buffer prior to addition of the tissue, or are added to the tissue if longer incubation is desired. Specific ion uptke is the difference between total uptake and that obtained with Naja α-toxin-treated Torpedo membranes.

Interactions of [^{3}H]H$_{12}$-HTX and [^{3}H]PCP with nACh receptors of Torpedo

The apparent rate of binding of [^{3}H]H$_{12}$-HTX or [^{3}H]PCP to nACh receptors of Torpedo in absence of receptor activation is quite slow; equilibrium is reached after 2-4 hours. In presence of ACh, the initial rate of binding of both ligands is accelerated in a dose-dependent manner as shown for [^{3}H]PCP in Fig. 1. At saturable concentrations of ACh, the apparent rate increased 2-3 orders of magnitude in 30 sec exposure. Pretreatment of Torpedo membranes with α-BGT inhibited totally the ACh-induced increase in binding, but [^{3}H]H$_{12}$-HTX or [^{3}H]PCP continued to bind slowly with the same rate as binding to resting nACH receptors.

It is important to note that binding of [^{3}H]H$_{12}$-HTX or [^{3}H]PCP is stimulation dependent, thus may be used as a measure of receptor activation. The uptake of ^{22}Na$^+$ into Torpedo microsacs increased as a function of carbamylcholine concentration, which gave a dose-response curve whose ED$_{50}$ agreed well with he known K$_d$ of carbamyl-choline on Torpedo receptors (Fig. 2). The effect of carbamylcholine on the initial rate of binding of [^{3}H]H$_{12}$-HTX corresponded with carbamylcholine activation of ^{22}Na$^+$ flux, suggesting that receptor activation was responsible for both parameters. Pretreatment of Torpedo membranes with α-BGT (1 μM) for 30 min completely inhibited the carbamylcholine-stimulated ^{22}Na$^+$ uptake without affecting [^{3}H]H$_{12}$-HTX binding, and H$_{12}$-HTX inhibited receptor-induced ^{22}Na$^+$ uptake in a dose-dependent manner without affecting equilibrium binding of [^{3}H]ACh (Fig. 3).

Receptor desensitization induced by preincubation of Torpedo membranes with agonists (e.g., ACh) resulted in a diminished ability of carbamylcholine to stimulate the initial rate of [^{3}H]H$_{12}$-HTX

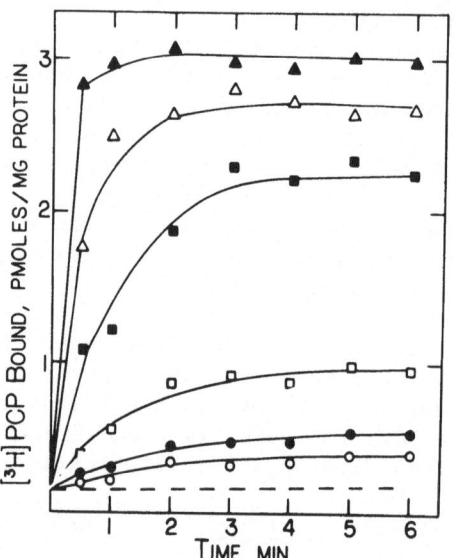

Fig. 1. Influence of ACh on the initial rate of binding of [^{3}H]PCP
(2 nM) to <u>Torpedo</u> nACh receptors. No ACh (O) or in presence
of ACh, 1 nM (●), 10 nM (□), 30 nM (■), 100 nM (△), and 300
nM (▲). (From Eldefrawi et al., 1980c).

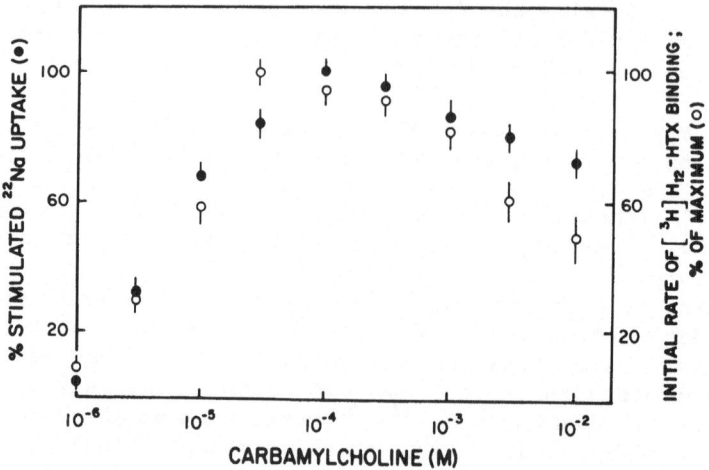

Fig. 2. Correlation between H$_{12}$-HTX binding to the ionic channels of
nACh receptors and ^{22}Na uptake of <u>Torpedo</u> membranes. The
effect of carbamylcholine on binding of 2 nM [^{3}H]H$_{12}$-HTX and
receptor-regulated ^{22}Na$^+$ uptake. (From Eldefrawi et al.,
1980b).

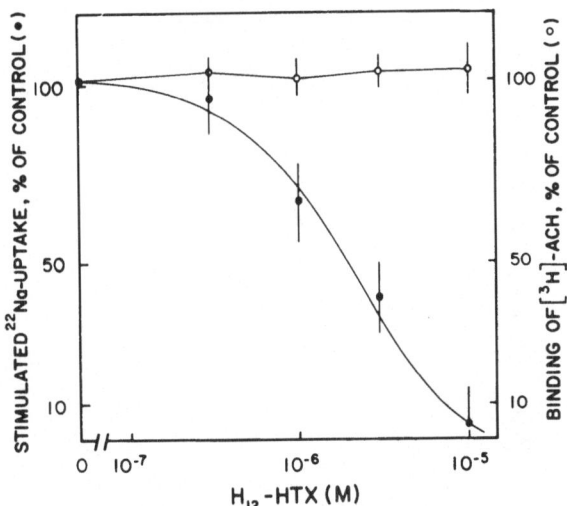

Fig. 3. The effect of H_{12}-HTX on receptor-activated (100 µM carba-
mycholine) $^{22}Na^+$ uptake and binding of 0.1 µM [^{3}H]ACh to
<u>Torpedo</u> nACh receptors. (From Eldefrawi et al., 1980b).

binding as compared with control preparations. The inhibition of the
100 µM carbamylcholine-stimulated [^{3}H]H_{12}-HTX binding had a rapid
phase that was complete by the earliest measurement (i.e., 6 sec) and
a slow one ($t_{1/2}$ 55 sec) which reached a plateau after 120 sec. The
rapid phase was proposed to be due to desensitization since it was
dependent upon the concentration of agonist and temperature (el-
Fakahany et al., 1982). Therefore, the use of the allosteric probes
[^{3}H]H_{12}-HTX or [^{3}H]PCP allows identification of the different
receptor conformations.

<u>Nicotine</u> <u>as</u> <u>a</u> <u>depolarizing</u> <u>blocker</u> <u>of</u> <u>nACh</u> <u>receptors.</u>

The stimulant and depressant actions of nicotine on the
mammalian autonomic nervous system were first described by Langley
(1901). At low concentrations (10^{-6}-10^{-4} M), nicotine potentiated
the initial rate of [^{3}H]H_{12}-HTX binding to the channel sites of nACh
receptors in <u>Torpedo</u> membrnes, but at >10^{-4} M nicotine the potentiation was inhibited (Fig. 4). The potentiation was due to the binding
of nicotine to the receptor sites (Aziz and Eldefrawi, 1973) and
action as an agonist. When the experiment was repeated with <u>Torpedo</u>
membranes pretreated with <u>Naja</u> α-neurotoxin, nicotine could not bind
to the inhibited receptor sites, and the potentiation of [^{3}H]H_{12}-HTX
binding was cancelled. However, >10^{-4} M nicotine inhibited [^{3}H]H_{12}-
HTX binding, suggesting direct binding to the channel sites. Thus,
the well known biphasic action of nicotine on insect ganglia (Flattum
and Shankland, 1971) and vertebrate autonomic ganglia and skeletal
muscle may be due to its action as an agonist at low concentration

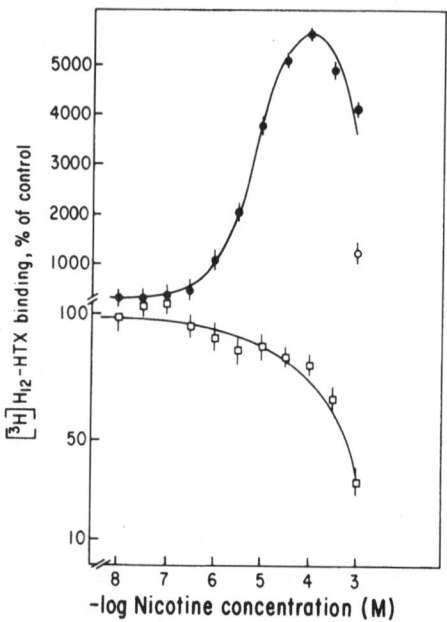

Fig. 4.   Effect of nicotine on the binding of [^{3}H]H$_{12}$-HTX to nACh
receptors of Torpedo. Untreated membranes were incubated
with 2 nM [^{3}H]H$_{12}$-HTX for 30 sec (O), or preincubated with
nicotine for 30 min prior to incubation with [^{3}H]H$_{12}$-HTX for
30 sec (●), or α-BGT-treated membranes (10 μM) for 60 min,
were incubated with 2 nM [^{3}H]H$_{12}$-HTX and nicotine for 120
min (□). (From Eldefrawi et al., 1982b).

and inhibitor of the receptor's channel at high concentrations as
well as inducer of receptor desensitization as shown by the reduction
of potentiation of [^{3}H]H$_{12}$-HTX binding upon preincubation with
nicotine (Fig. 4).

The effect of nicotine on binding of ligands to house fly nACH receptors

α-BGT inhibits nACh receptor function in the cockroach CNS
(Harrow and Sattelle, 1983). A nACh receptor was identified in house
fly brain membrane by virtue of its specific binding of [^{125}I] α-BGT
(Eldefrawi and Eldefrawi, 1980). Its K_d of 3nM compares with K_d
values of 1.1-1.8 nM for comparable binding to Drosophila membranes
(Schmidt-Nielsen et al., 1977; Rudloff, 1978). These are lower
affinities than for the nACh receptor of Torpedo electric organ.
Although the two receptors have nicotinic drug specificity, that of
the house fly brain has much lower affinity for decamethonium (Table
1), which is also a poor inhibitor of α-BGT binding to rat brain
membranes (McQuarrie et al., 1976).

Table. 1. Drug profiles of nACh receptors of house fly brain and
           Torpedo electric organ (from Eldefrawi and Eldefrawi,
           1980)

(10 µM)	% Inhibition of $[^{125}I]$ α-BGT binding	
	House fly head[1]	Torpedo electroplax
ACH	37	70
d-Tubocurarine	51	63
Decamethonium	0	64
Atropine	0	5
Scopolamine	0	7

[1] $[^{125}I]$ α-BGT used at 5 nM.

Nicotine inhibited binding of $[^{125}I]$ α-BGT to house fly brain
membranes (Fig. 5). Since preincubation with nicotine caused a shift
to the left of the dose-response function (Fig. 6), it suggested that
the increased potency of nicotine to inhibit $[^{125}I]$ α-BGT binding was
due to increased affinity for nicotine by the desensitized receptor
conformation.

Since HTX blocks ACh responses in insect neurons in the same
manner it affects nACh receptors of vertebrates (Artola et al.,
1984), one would have expected to see stimulation of $[^3H]H_{12}$-HTX
binding by nicotine to the house fly preparation as well. Unfortunately, we could not study the effect of nicotine on the ionic channel
of the house fly nACh receptor, because the house fly head preparation had very high $[^3H]H_{12}$-HTX nonspecific binding (Eldefrawi and
Eldefrawi, 1982). Thus, any modulation of specific $[^3H]H_{12}$-HTX
binding (representing 2% of total binding) by an agonist was masked
by the large amount of nonspecific binding. However, successful
manipulation of house fly brain preparation to reduce the numbers of
nonspecific sites and increase the relative density of $[^3H]H_{12}$-HTX
specific binding sites is possible. When this is accomplished we
shall have a powerful assay to study interactions of insecticides
with nACh receptors in insect neural tissue.

Anticholinesterases as receptor agonists and noncompetitive blockers

Structural activity relationship studies are always halpful in
understanding the mechanism of action of drugs. We studied the
interactions of six symmetrically substituted tetraalkylammonium (C_1-
C_6) anticholinesterases with the nACh receptor of Torpedo electric
organ (Bakry et al., 1982). Tetramethylammonium (TMA) was a rela-

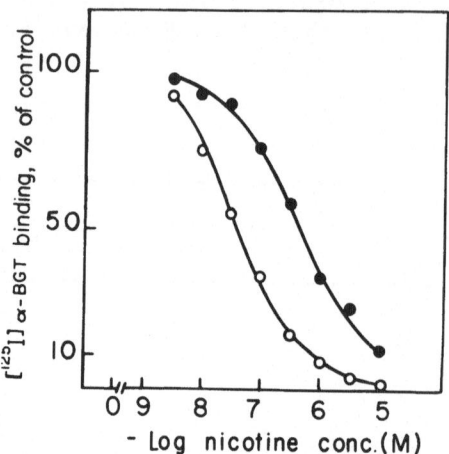

Fig. 5. The effect of nicotine on the initial rate of binding of 1
nM [^{125}I] α-BGT to house fly head membranes. Nicotine was
added 10 sec prior to incubation with α-BGT for 40 sec (●)
or preincubated with tissue for 30 min prior to incubation
with α-BGT for 40 sec (O).

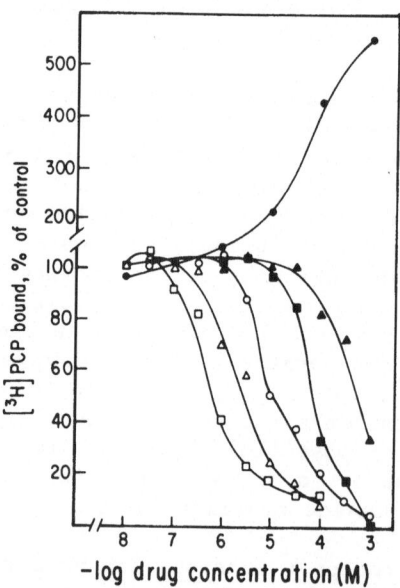

Fig. 6. The effect of n-tetraalkylammonium (C$_1$-C$_6$) compounds on the
binding of [^{3}H]PCP to the channel sites of <u>Torpedo</u> nACh
receptors. Incubation with 2 nM [^{3}H]PCP was for 30 sec in
absence of carbamylcholine and presence of C$_1$ (●), C$_2$ (▲),
C$_3$ (■), C$_5$ (△), C$_6$ (□). (From Bakry et al., 1982).

tively potent inhibitor of [^{3}H]ACh binding (K_i = 38 µM), tetra-
ethylammonium (TEA) was weaker (K_i 200 µM), but none of the other
four longer-chain compounds up to 1 mM inhibited [^{3}H:ACh binding.
This suggested that only TMA and TEA bound to the receptor. However,
all compounds except TMA inhibited binding of [^{3}H]-PCP to the receptor's channel sites (Fig. 6), thus acting like channel blockers. TMA
stimulated the binding like an agonist. The data suggest that
quaternary ammonium compounds may bind to both allosteric and receptor sites, and increasing the lipid solubility and probably size
converts the compound from an agonist to a channel blocker. It is
interesting to note that such changes in structure also increase
permeability of quarternary ammonium compounds into the insect CNS
(Eledfrawi and O'Brien, 1967). It will be interesting to know the
relative insecticidal potency of this homologous series of quaternary
ammonium compounds.

    Carbamate and organophosphate anticholinesterases also bound to
nACh receptors and modified their responses. Neostigmine, pyridosti-
gmine and physostigmine inhibited binding of the receptor ligands
[^{3}H]ACh and [^{125}I] α-BGT. In addition, preexposure of Torpedo
membranes to the carbamates enhanced the receptor's affinity for
them, and their potency in displacing [^{125}I] α-BGT binding (Fig. 7),
suggesting that they act as agonists and cause receptor desensitization. The carbamates, except for physostigmine, enhanced [^{3}H]H$_{12}$-HTX
binding like partial agonists, since they bound to the receptor site,

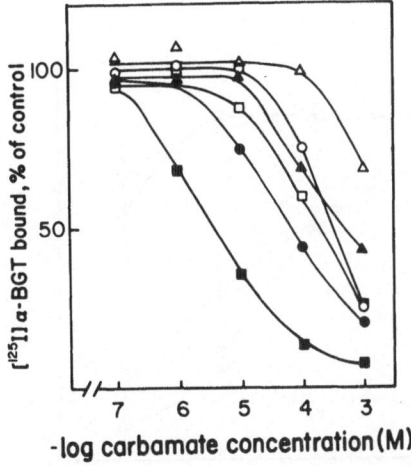

Fig. 7. Inhibition of the initial rate of binding of [^{125}I] α-BGT
         (1 nM) to nACH receptors of Torpedo by three carbamate
         anti-ACh-esterases following 10 sec (open symbols) or 30 min
         (solid symbols) exposure prior to incubation with [I^{125}]
         α-BGT for 40 sec. Neostigmine (□,■), physostigmine (O,●)
         and pyridostigmine (△,▲).

though at high concentrations. Physostigmine did not potentiate much [^{3}H]H$_{12}$-HTX binding because it also bound to the channel sites (Fig. 8) and acted as a channel blocker.

Modulation of nACh receptors by pyrethroids

Pyrethroids were divided into two types based on the symptoms they produced in the American cockroach (Gammon et al., 1981) and the rat (Aldridge, 1980) as well as on their effects on the cockroach cercal sensory nerve (Gammon et al., 1981) and the crayfish giant axon (Narahashi, 1980). Type I pyrethroids cause restlessness, hyperactivity, and paralysis in the cockroach, as well as repetitive nerve firing in its cercal sensory nerves (Gammon et al., 1981). On the other hand, type II pyrethroids produce a convulsive phase and do not cause repetitive firing in these nerves, although they cause bursts of spikes in the cercal motor nerve. Type II compounds contain an α-cyano substituent. However, there are a few exceptions to this division, where an α-cyano pyrethroid produces type I action on the cockroach cercal sensory nerve but type II symptoms in the cockroach (Gammon et al., 1981) or type I symptoms in the rat Also, type II pyrethroids have a stronger excitatory effect than type I on crayfish sensory nerve, but the reverse potency on cockroach cercal nerve cord, and some α-cyano pyrethroids produce a depolarizing block of crayfish giant axon without repetitive firing (Narahashi, 1980; Gammon et al., 1981).

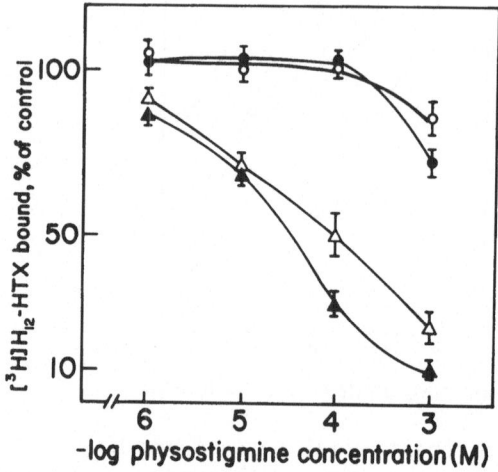

Fig. 8. The effect of physostigmine (●,▲) and physostigmine methiodide (○,△) on the equilibrium binding of 2 nM [^{3}H]H$_{12}$-HTX (120 min) to nACh receptors of <u>Torpedo</u> in absence (○,●) or presence (△,▲) of 100 μM carbamylcholine. (Aldridge, 1980).

We found that the Torpedo nACh receptor interacted with pyrethroids as shown by the noncompetitive inhibition of the ACh-induced $[^3H]H_{12}$-HTX binding to Torpedo receptors by allethrin (Fig. 9) and 18 other pyrethroids (Abbassy et al., 1983a,b). Based on the speed of action on this receptor, the pyrethroids were also divided into two types (Fig. 10). Fast-acting ones that produced maximal inhibition within 30 sec (type I), and are chrysanthemic esters of cyclopentenolone except for tetramethrin and resmethrin, which are esters of tetrahydrophthalimidomethyl and 5-benzyl-3-furylmethyl alcohols, respectively. Slow acting ones produced maximal inhibition after 1 h (type II), and are esters of 3-phenoxybenzyl alcohol containing the α-cyano substituent, except for permethrin which does not. The difference in the kinetics of action of the two types is illustrted in the effects of allethrin (type I) and fluvalinate (type II) on the kinetics of $[^3H]H_{12}$-HTX binding (Fig. 11). In general, these two types correspond to the ones divided on the basis of symptoms or actions on the nerve.

Pyrethroids were more potent inhibitors of $[^3H]H_{12}$-HTX on the nicotinic ACh receptor at lower temperatures because of increased affinity (Fig. 12). The negative temperature coefficient of pyrethroids in their blockade of the nACh receptor is in line with their higher toxicity at lower temperatures (Narahashi, 1971), which may be due in part to reduced pyrethroid metabolism at lower temperatures. However, nerve blocking action of pyrethroids also has a negative temperature coefficient through greater suppression of sodium conductance at lower temperatures (Wang et al., 1972). On the other hand, there are conflicting reports on the effects of tempera-

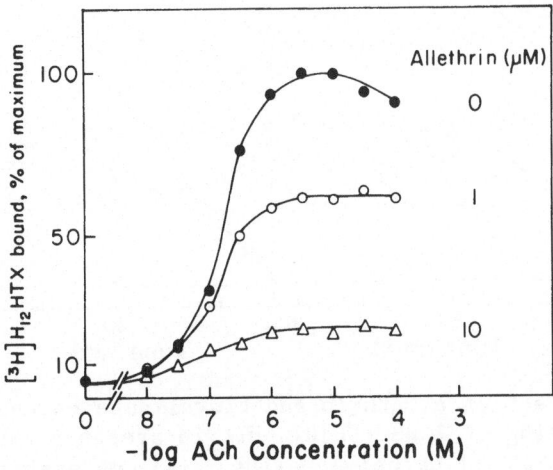

Fig. 9. Effect of allethrin on the binding of 2 nM $[^3H]H_{12}$-HTX to nACh receptors of Torpedo in the presence of varying concentrations of ACh and allethrin: zero M (●), 1 μM (o) and 10 μM (Δ). (From Abbassy et al., 1983a).

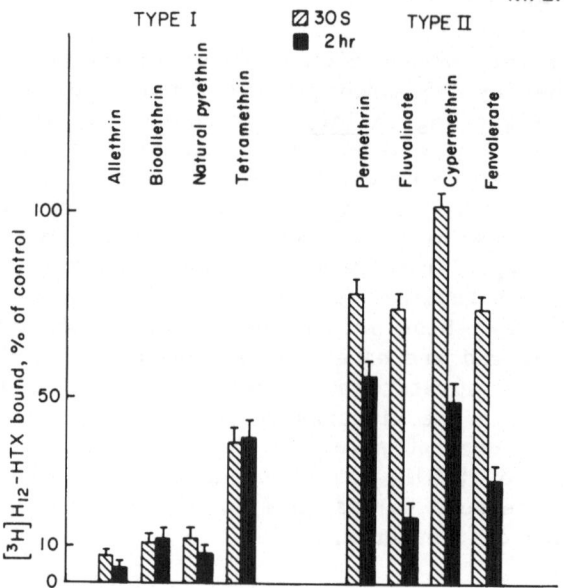

Fig. 10.   Effect of pyrethroids on binding of 2 nm $[^3H]H_{12}$-HTX in
           presence of 100 μM carbamylcholine measured after 30 sec
           (stripe) and 2 hours (solid).   All pyrethroids were present
           at 100 μM final concentration.   (From Abbassy et al.,
           1983b).

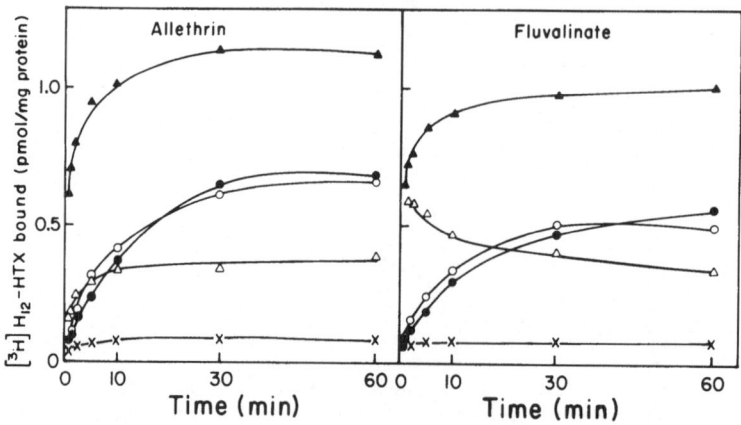

Fig. 11.   The effect of allethrin and fluvalinate on the time course
           of binding of 2 nM $[^3H]H_{12}$-HTX in presence (Δ) or absence
           (O) of 100 μM carbamylcholine.   (▲) represents binding in
           presence of 100 μM carbamylcholine but no pyrethroids, and
           (●) binding in absence of pyrethroids and carbamylcholine.
           (x) represents binding in presence of 5 mM of the channel
           blocker amantadine.   (From Abbassy et al., 1983b).

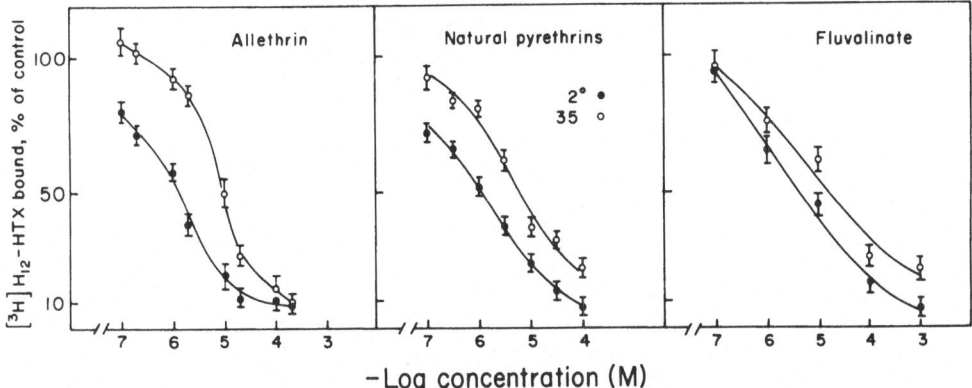

Fig. 12. The effect of temperature on the potency of pyrethroids to displace binding of 2 nM [^{3}H]H$_{12}$-HTX to Torpedo receptors in presence of 100 M carbamylcholine. (From Abbasy et al., 1983b).

ture on repetitive nerve activity; it occurs only at high temperature in the cockroach giant axon (Narahashi, 1962), at medium temperature in squid giant axon (Narahashi, 1962), is unaffected by temperatures in frog peripheral nerves (van den Bercken et al., 1973), or is increased greatly at low temperatures in the lateral-line organ of Xenopus (van den Bercken et al., 1973) and frog motor endplate (Wouters et al., 1977), lateral-line sense organ, and motor nerve terminal (van den Bercken, 1977).

Even though pyrethroids affected binding of ligands to the nACh receptor, they were reported not to inhibit vertebrate endplate potentials (Evans, 1976; Wouters et al., 1977), possibly because of the masking by their potent presynaptic excitatory effect due to action of Na$^+$ channels. We attempted to study biochemically their effect on receptor function by monitoring receptor-activated ^{45}Ca^{2+} flux in Torpedo microsacs, and found that they inhibited it (Fig. 13). Initially, this ^{45}Ca^{2+} uptake was believed to be a Ca^{2+} fluxing through the receptor/channel across the membrane. However, in more recent experiments where the effects of allethrin and fluvalinate were compared on receptor-regulated ^{22}Na$^+$ and ^{45}Ca^{2+} uptake by Torpedo membrane, the pyrethroids selectively inhibited ^{45}Ca^{2+} uptake (Fig. 14). This raises many interesting questions on whether Ca^{2+} uptake represents Ca^{2+} binding to a membrane component associated with receptor activation and whether or not this component is a part of the nACh receptor molecule. Answering these and other questions should help us understand the effect of pyrethroids on synaptic transmission in general and on cholinergic transmission in particular.

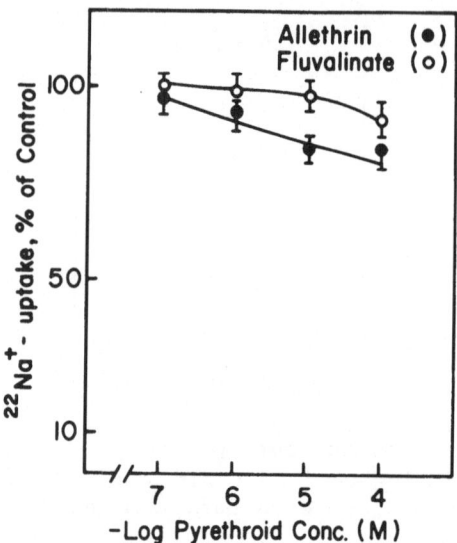

Fig. 13.   The effect of allethrin (●) and fluvalinate (○) on the
           receptor-regulated $^{22}Na^+$ uptake. Specific uptake was that
           measured in presence of 100 μM carbamylcholine and
           inhibited by pretreatment with <u>Naja</u> α-neurotoxin.
           Pyrethroids were incubated with the tissue for 15 min at
           4°C prior to assay.

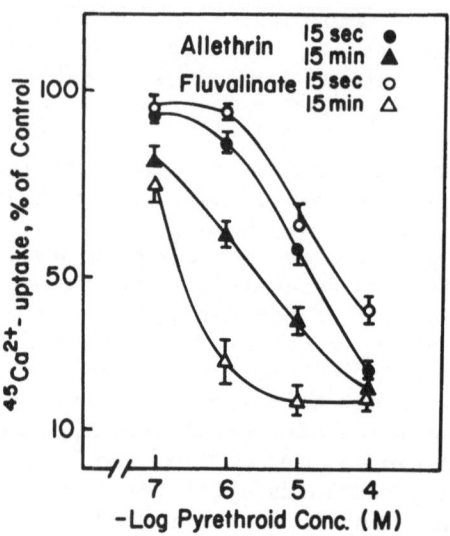

Fig. 14.   The effect of allethrin (●,▲) and fluvalinate (○,△) on the
           receptor-regulated $^{45}Ca^{2+}$ uptake. Pyrethroids were added
           to the tissue either 15 sec (●,○) or 30 min (▲,△) prior
           to addition of $^{45}Ca^{2+}$ and carbamylcholine.

DISCUSSION

Pyrethroids cause hyperexcitability and also blockade of nerve action. The latter may be a secondary consequence of pyrethroid poisoning, since much higher concentrations may be required for blockade than for hyperexcitability. Also, the cockroach cercal afferent neurons are hyperexcited soon after exposure to allethrin, but are not blocked till hours after paralysis (Gammon, 1978). In addition, conduction in its giant fibers is unimpaired even when the cockroach is showing severe symptoms of poisoning, yet the sixth abdominal ganglion has increased activity (Burt and Goodchild, 1971). The relationship of repetitive firing in axons to pyrethroid toxicity is also questioned based on the finding that certain pyrethroids cause it and others do not (Narahashi, 1980; Gammon et al., 1981). Further, pyrethroids that induce repetitive activity in locust sensory fibers do not induce it in motor axons, and _vice versa_ (Clements and May, 1977). It appears that although the Na^+ channel may be the primary target for pyrethroid action, they may act on other targets as well, such as the nACh receptor (Abbassy et al., 1982) and the GABA receptor (Gammon et al., 1982; Abalis et al., 1983). Thus, multiple targets may be involved in the knockdown and toxic actions of pyrethroids.

In conclusion, it is evident that several kinds of insecticides ranging from anticholinesterases to pyrethroids bind to and affect the function of nACh receptors in insects or vertebrates. Whether the nACh receptor is a primry or secondary molecular target for the toxic actions of insecticides depends on the receptor's affinity for the compound and the mechanism of its action. There are reported differences between insect and vertebrate receptors, and undoubtedly further studies will uncover more specific differences. Such differences in receptor pharmacology between insects and mammals may be exploited to develop insecticides with low mammalian toxicity.

ACKNOWLEDGMENT

Our reported research was financed by NIH grant ES 02594 and Army Research Office grant DAAG 29-81-0161. We thank our secretary, Evelyn Elizabeth, for her excellent typing.

REFERENCES

Abbassy, M.A., Eldefrawi, M.E., and Eldefrawi, A.T., 1982, Allethrin interactions with the nicotinic acetylcholine receptor channel, Life Sci., 31:1547-1552.
Abbassy, M.A., Eldefrawi, M.E., and Eldefrawi, M.D., 1983a, Pyrethroid action on the nicotinic receptor/channel, Pestic. Biochem. Physiol., 19:299-308.

Abbassy, M.A., Eldefrawi, M.E., and Eldefrawi, A.T., 1983b, Influence of the alcohol moiety of pyrethroids on their interactions with the nicotinic acetylcholine receptors, J. Toxicol. Environ. Health, 12:575-590.

Albuquerque, E.X., Eldefrawi, A.T., Eldefrawi, M.E., Mansour, N.A., and Tsai, M.-C., 1978, Amantadine: neuromuscular blockade by suppression of ionic conductance of the acetylcholine receptor, Science, 199:788-790.

Albuquerque, E.X., Tsai, M.-C., Aronstam, R.S., Witkop, B., Eldefrawi, A.T., and Eldefrawi, M.E., 1980, Phencyclidine interactions with ionic channels of the acetylcholine receptor and electrogenic membranes, Proc. Natl. Acad. Sci. USA, 77:1224-1228.

Aronstam, R.S., Eldefrawi, A.T., Pessah, I.N., Daly, J.W., Albuquerque, E.X., and Eldefrawi, M.E., 1981, Regulation of [^{3}H]perhydrohistrionicotoxin binding to Torpedo electroplax by effectors of the acetylcholine receptor, J. Biol. Chem., 73:440-450.

Artola, A., Callec, J.J., Hue, B., David, J.A., and Sattelle, B.D., 1984, Actions of amantadine at synaptic and extrasynaptic cholinergic receptors in the central nervous system of the cockroach Periplaneta americana, J. Insect Physiol., 30:185-190.

Aziz, S.A., and Eldefrawi, M.E., 1973, Cholinergic receptors of the central nervous system of insects, Pestic. Biochem. Physiol., 3:168-174.

Bakry, N.M., Eldefrawi, A.T., Eldefrawi, M.E. and Riker, W.F., Jr., 1982, Interactions of quaternary ammonium drugs with acetylcholinesterase and acetylcholine receptor of Torpedo electric organ, Mol Pharmacol., 22:63-71.

Bettini, S., D'Ajello, V., and Maroli, M., 1973, Cartap activity on the cockroach neurons and neuromuscular transmission, Pestic. Biochem. Physiol., 3:199-205.

Birdsall, N.J.M. and Hulme, E.C., 1976, Biochemical studies on muscarinic acetylcholine receptors, J. Neurochem., 27:7-16.

Blanchard, S.G., Elliott, J., and Raftery, M.A., 1979, Interaction of local anesthetics with Torpedo californica membrane-bound acetylcholine receptor, Biochemistry, 18:5880-5885.

Bradley, R.J., Howell, J.H., Romine, W.O., Carl, C.F., and Kemp, G.E., 1976, Characterization of a nicotine acetylcholine receptor from rabbit skeletal muscle and reconstitution in planar phospholipid bilayers, Biochem. Biophys. Res. Comm., 68:577-584.

Burt, P.E. and Goodchild, R.E., 1971, The site of action of pyrethrin I in the nervous system of the cockroach Periplaneta americana, Entomologia Exp. Appl., 14:179-189.

Clements, A.N. and May, T.E., 1977, The actions of pyrethroids upon the peripheral nervous system and associated organs in the locust, Pestic. Sci., 8:661-680.

Dolly, J.O, Albuquerque, E.X., Sarvey, J.M., Mallick, B., and Barnard, E.A., 1976, Binding of perhydrohistrionicotoxin to the postsynaptic membrane of skeletal muscle in relation to its blockade of acetylcholine-induced depolarization, Mol. Pharmacol., 13:1-14.

Dudai, Y., 1978, Properties of an α-bungarotoxin-binding cholinergic nicotinic receptor from Drosophila melanogaster, Biochem. Biophys. Acta, 539:505-517.

Dudai, Y., and Ben-Barak, J., 1977, Muscarinic receptor in Drosophila melanogaster demonstrated by binding of [^{3}H]quinuclidinyl benzilate, FEBS Letters, 81:134-136.

Eldefrawi, M.E., Aronstam, R.S., Bakry, N.M., Eldefrawi, A.T., and Albuquerque, E.X., 1980b, Activation, inactivation and desensitization of acetylcholine receptor channel complex detected by binding of perhydrohistrionicotoxin, Proc. Natl. Acad. Sci. USA, 77:2309-2313.

Eldefrawi, A.T., Bakry, N.M., Eldefrawi, M.E., Tsai, M.-C., and Albuquerque, E.X., 1980a, Nereistoxin interaction with the acetylcholine receptor-ionic channel complex, Mol. Pharmacol, 17:172-179.

Eldefrawi, M.E., Britten, A.G., and O'Brien, R.D., 1971, Action of organophosphates on acetylcholine receptors, Pestic. Biochem. Physiol., 1:101-108.

Eldefrawi, M.E., and Eldefrawi, A.T., 1973, Purification and molecular properties of the acetylcholine receptor from Torpedo electroplax, Arch. Biochem. Biophys., 159:362-373.

Eldefrawi, M.E., and Eldefrawi, A.T., 1980, Putative acetylcholine receptors in housefly brain, In, "Receptors for Neurotransmitters, Hormones and Pheromones in Insects," D.B. Sattelle, L.M. Hall, and J.G. Hildebrand, eds., Elsevier/North-Holland, Amsterdam.

Eldefrawi, M.E., and Eldefrawi, A.T., 1983, Neurotransmitter receptors as targets for pesticides, J. Environ. Sci. Health, B18:65-88.

Eldefrawi, A.T., Eldefrawi, M.E., Albuquerque, E.X., Oliveiri, A.C., Mansour, N.A., Adler, M., Daly, J.W., Brown, G.B., Burgermeister, W., and Witkop, B., 1977, Perhydrohistrionicotoxin: a potential ligand for the ion conductance modulator of the acetylcholine receptor, Proc. Natl. Acad. Sci. USA, 74:2172-2176.

Eldefrawi, M.E., Eldefrawi, A.T., Aronstam, R.S., Maleque, M.A., Warnick, J.E., and Albuquerque, E.X., 1980c, [^{3}H]Phencyclidine: a probe for the ionic channel of the nicotinic receptor, Proc. Natl. Acad. Sci. USA, 77:7458-7462.

Eldefrawi, M.E., Eldefrawi,, A.T., and O'Brien, R.D., 1970, Mode of action of nicotine in the housefly, J. Agr. Food Chem., 18:1113-1116.

Eldefrawi, M.E. and Eldefrawi, A.T., and O'Brien, R.D., 1971a, Multiple binding sites for cholinergic ligands in a particulate fraction of Electrophorus electroplax, Proc. Natl. Acad. Sci. USA, 68:1047-1050.

Eldefrawi, M.E., Eldefrawi, A.T. and O'Brien, R.D., 1971b, Binding of five cholinergic ligands to housefly brain and Torpedo electroplax: Relationship to acetylcholine receptors, Mol. Pharmacol., 7:104-110.

Eldefrawi, A.T., Miller, E.R., and Eldefrawi, M.E., 1982b, Binding of depolarizing drugs to the ionic channel sites of the nicotinic acetylcholine receptor, Biochem. Pharmacol., 31:1819-1822.

Eldefrawi, M.E., and O'Brien, R.D., 1967, Permeability of the abdominal nerve cord of the American cockroach, Periplaneta americana L., to quaternary ammonium salts, J. Exp. Biol. 46:1-12.

Eldefrawi, A.T., Shaker, N., and Eldefrawi, M.E., 1982a, Binding of acetylcholine receptor/channel probes to housefly head membranes, In, "Neuropharmacology of Insects," Ciba Found. Symp. 88, Pitman, London.

El-Fakahany, E.F., Eldefrawi, A.T., and Eldefrawi, M.E., 1982, Nicotinic acetylcholine receptor desensitization studied by [^{3}H]perhydrohistrionicotoxin binding, J. Pharmacol. Exp. Ther., 221:694-700.

Epstein, M., and Racker, E., 1978, Reconstitution of carbamylcholine-dependent sodium ion flux and desensitization of the acetylcholine receptor from Torpedo californica, J. Biol. Chem., 253:6660-6662.

Evans, M.H., 1976, End-plate potentials in frog muscle exposed to a synthetic pyrethroid, Pestic. Biochem. Physiol., 6:547-550.

Flattum, R.F. and Shankland, D.L., 1971, Acetylcholine receptors and the diphasic action of nicotine in the American cockroach Periplaneta americana (L.), Comp. Gen. Pharmacol., 2:159-167.

Gammon, D.W., 1978, Neural effects of allethrin on the free walking cockroach Periplaneta americana: and investigation using defined doses at 15 and 32°C, Pestic. Sci., 9:79-91.

Gammon, D.W., Brown, M.A., and Casida, J.E., 1981, Two classes of pyrethroid action in the cockroach, Pestic. Biochem. Physiol., 15:181-191.

Gammon, D.W., Lawrence, L.J., and Casida, J.E., 1982, Pyrethroid toxicology: protective effects of diazepam and phenobarbital in the mouse and the cockroach, Toxicol. Appl. Pharmacol., 66:290-296.

Haim, N., Nahum, S., and Dudai, Y., 1979, Properties of a putative muscarinic cholinergic receptor from Drosophila melanogaster, J. Neurochem., 32:543-552.

Harrow, I.D. and Sattelle, D.B., 1983, Acetylcholine receptors on the cell body membrane of giant interneurones in the cockroach Periplaneta americana, J. Exp. Biol., 105:339-350.

Hartig, P.R., and Raftery, M.A., 1977, Lactoperoxidase catalyzed membrane surface labeling of the acetylcholine receptor from Torpedo californica, Biochem. Biophys. Res. Comm., 78:16-22.

Heidmann, T., and Changeux, J.-P., 1978, Structural and functional properties of the acetylcholine receptor protein in its purified and membrane-bound sites, Ann. Rev. Biochem., 47:317-357.

Hess, G.P., Andrews, J.P., Struve, G.E., and Combs, S.E., 1975,
    Acetylcholine-receptor-mediated flux in electroplax membrane
    preparations, Proc. Natl. Acad. Sci. USA, 72:4371-4375.

Hildebrand, J.G., Hall, L.M., and Osmond, B.C., 1979, Distribution
    of binding sites for ^{125}I-labeled α-bungarotoxin in normal and
    deafferented antennal lobes of Manduca sexta, Proc. Natl. Acad.
    Sci. USA, 76:499-503.

Jones, S.W., Sudershan, P., and O'Brien, R.D., 1981, α-Bungarotoxin
    binding in house fly heads and Torpedo electroplax, J.
    Neurochem., 36:447-453.

Jones, S.W. and Sumikawa, K., 1981, Quinuclidinyl benzilate binding
    in housefly heads and rat brain, J. Neurochem, 36:454-459.

Karlin, A., 1980, Molecular properties of nicotinic acetylcholine
    receptors, In, "The Cell Surface and Neuronal Function," Cotman,
    C.W., Poste, G., and Nicolson, G.L. eds., Elsevier/North-Holland
    Biomedical Press, Amsterdam.

Kasai, M., and Changeux, J.-P., 1971a, In vitro excitation of puri-
    fied membrane fragments by cholinergic agonists II. The perme-
    ability change caused by cholinergic agonists, J. Membrane
    Biol., 6:26-57.

Kasai, M., and Changeux, J.-P., 1971b, In vitro excitation of puri-
    fied membrane fragments by cholinergic agonists III. Comparison
    of the dose-responsive curves to decamethonium with the corres-
    ponding binding curves of decamethonium to the cholinergic
    receptor, J. Membrane Biol., 6:58-80.

Katz, B., and Miledi, R., 1975, The effect of procaine on the action
    of acetylcholine at the neuromuscular junction, J. Physiol.
    (Lond), 230:707-717.

Kohanski, R.A., Andrews, J.P., Wins, P., Eldefrawi, M.E., and Hess,
    G.P., 1977, A simple quantitative assay of ^{125}I-labeled α-
    bugarotoxin binding to soluble and membrane-bound acetylcholine
    receptor protein, Anal. Biochem., 80:531-539.

Langley, J.N., 1901, On the stimulation and paralysis of nerve cells
    and of nerve-endings, J. Physiol. (Lond), 27:224-236.

Lindstrom, J., and Dau, P., 1980, Biology of myasthenia gravis, Ann.
    Rev. Pharmacol. Toxicol., 20:337-362.

Lowry, O.H., Rosebrough, N.J., Farr, A.L., and Randall, R.J., 1951,
    Protein measurement with the Folin phenol reagent, J. Biol.
    Chem., 193:265-275.

McQurrie, C., Salvaterra, P.M., De Blas, A., Routes, J., and Mahler,
    H.R., 1976, Studies on nicotinic acetylcholine receptors in
    mammalian brain, J. Biol. Chem., 251:6335-6339.

Miller, D.L., Moore, H.-P., Hartig, P.R., and Raftery, M.A., 1978,
    Fast cation flux from Torpedo californica membrane preparations:
    implications for a functional role for acetylcholine receptor
    dimers, Biochem. Biophys. Res. Comm., 85:632-640.

Narahashi, T., 1962, Effect of the insecticide allethrin on membrane
    potentials of cockroach giant axons, J. Cell. Comp. Physiol.,
    59:61-66.

Narahashi, T., 1971, Mode of action of pyrethroids, Bull. WHO, 44:337-345.

Narahashi, T., 1976, Effects of insecticide on nervous conduction and synaptic transmission, In, "Insecticide Biochemistry and Physiology," Wilkinson, C., ed., Plenum, New York.

Narahashi, T., 1980, Site and types of action of pyrethroids on nerve membrane, In, "Pyrethroid Insecticides: Chemistry and Action," Elliott, J.E. and Elliott, M., eds., Table Ronde Roussel-Uclaf, Paris, No. 37.

O'Brien, R.D., Eldefrawi, M.E., and Eldefrawi, A.T., 1972, Isolation of acetylcholine receptors, Ann. Rev. Pharmacol., 12:19-34.

Oswald, R., and Changeux, J.-P., 1981, Ultraviolet light-induced labeling by noncompetitive blockers of the acetylcholine receptor from Torpedo marmorata., Proc. Natl. Acad. Sci. USA, 78:3925-3929.

Raftery, M.A., Hunkapiller, M.W., Strader, C.D., and Hood, L.E., 1980, Acetylcholine receptor: complex of homologous subunits, Science, 208:1454-1457.

Rudloff, E., 1978, Acetylcholine receptors in the central nervous system of Drosophila melanogaster, Exp. Cell Res., 111:185-190.

Sakai, M., 1964, Studies on the insecticidal action of nereistoxin, 4-N,N-dimethylamino-1,2-dithiolane I. Insecticidal properties, Jap. J. Appl. Ent. Zool., 8:324-333.

Sakai, M., 1966, Studies on the insecticidal action of nereistoxin, 4-N,N-dimethylamino-1,2-dithiolne IV. Role of the anticholinesterase activity in the insecticidal action to housefly, Musca domestica (Diptera:Muscidae), Appl. Entomol. Zool., 1:73-82.

Sakai, M., 1967, Studies on the insecticidal action of nereistoxin, 4-N,N-dimethylamino-1,2-dithiolane V. Blocking action on the cockroach ganglion, Botyu-Kagaku, 32:21-33.

Sattelle, D.B., and David, J.A., 1983, Volatage-dependent block by histrionicotoxin of the acetylcholine-induced current in an insect motoneurone cell body, Neurosci. Lett., 43:37-41.

Sattelle, D.B., David, J.A., Harrow, I.D., and Hue, B., 1980, Actions of α-bungarotoxin on identified insect central neurones, In, "Receptors for Neurotransmitters, Hormones, and Pheromones in Insects," Sattelle, D.B., Hall, L.M., and Hildebrand, J.G., eds, Elsevier/North-Holland, Amsterdam.

Schmidt, J., and Raftery, M.A., 1973, A simple assay for the study of solubilized acetylcholine receptors, Anal. Biochem., 52:349-354.

Schmidt-Nielsen, B.K., Gepner, J.I., Teng, N.N.H., and Hall, L.M., 1977, Characterization of an α-bungrotoxin binding component from Drosophila melanogaster, J. Neurochem, 29:1013-1029.

Shaker, N., and Eldefrawi, A.T., 1981, Muscarinic receptor in housefly brain and its interaction with chlorobenzilate, Pestic. Biochem. Physiol., 15:14-20.

Shankland, D.L., Rose, J.A., and Donniger, C., 1971, The cholinergic nature of the cercal nerve-giant fiber synapse in the six abdominal ganglion of the American cockroach, J. Neurobiol.,

2:247-262.

Strader, C.D. and Raftery, M.A., 1980, Topographic studies of Torpedo acetylcholine receptor subunits as a transmembrane complex, Proc. Natl. Acad. Sci. USA, 77:5807-5811.

Van Den Bercken, J., Akkermans, L.M.A., and Zalm, J.M., 1973, DDT-like action of allethrin in the sensory nervous system of Xonopus laevis, Eur. J. Pharmacol., 21:95-106.

Van Den Bercken, J., 1977, The action of allethrin on the peripheral nervous system of the frog, Pestic. Sci., 8:692-699.

Wang, C.M., Narahashi, T., and Scuka, M., 1972, Mechanism of negative temperature coefficient of nerve blocking action of allethrin, J. Phrmacol. Exp. Ther., 182:442-454.

Weiland, G., Georgia, B., Wee, V.T., Chignell, C.F., and Taylor, P., 1976, Ligand interactions with cholinergic receptor-enriched membranes from Torpedo: influence of agonists exposure on receptor properties, Mol. Pharmacol., 12:1091-1105.

Wouters, W., Van Den Bercken, J., and Van Ginneken, A., 1977, Presynaptic action of the pyrethroid insecticide allethrin in the frog motor end plate, Eur. J. Pharmacol., 43:163-171.

Yamamoto, I., 1965, Nicotinoids as insecticides, Adv. Pest Control Res., 6:231-260.

Young, A.P., and Sigman, D.S., 1981, Allosteric effects of volatile anesthetics on the membrane-bound acetylcholine receptor protein 1. Stabilization of the high affinity state, Mol. Pharmacol., 20:498-505.

THE ACTION OF GLUTAMATE AGONISTS AT INSECT NEUROMUSCULAR JUNCTION

Jun-Ichi Fukami

Laboratory of Insect Toxicology
Institute of Physical and Chemical Research
Wako-Shi, Saitama, 351-01, Japan

INTRODUCTION

The effects of glutamate agonists and some neuro-active compounds at the neuromuscular junction of the mealworm, Tenebrio molitor, were studied. Glutamate agonists (L-glutamic acid and quisqualic acid) produced membrane depolarization accompanied with the
suppression of the amplitude of EPSPs. IC_{50} values were 2.3 x 10^{-5}M
for quisqualic acid and 4.2 x 10^{-5}M for L-glutamic acid. These two
agonists were most effective among compounds tested in this study.

One hundred twenty analogues of quisqualic acid were synthesized
and assayed. Two new agonists for amino acid receptors, L-glutamic
acid N-thiocarboxyanhydride (L-GANTA) and DL-hydantoinpropionic acid
(DL-HPA), were discovered in this study. L-GANTA and DL-HPA produced
muscle membrane depolarization, accompanied with a reduction of
muscle input resistance. The amplitude of EPSPs decreased in the
presence and absence of L-GANTA and DL-HPA. The apparent dissociation constants (K_d) obtained from dose-depolarization plots were 7 x
10^{-4}M for L-GANTA and 9 x 10^{-4}M for DL-HPA. Some structural requirements for active agonists to interact at the amino acid receptors on
insect muscle were discussed.

REVIEW OF RELATED WORK

L-Glutamic acid is a leading candidate for the excitatory transmitter at the neuromuscular junction of insects (Usherwood, 1978;
Nistri and Constanti, 1979). L-Glutamic acid, when applied to the
bathing medium for insect neuromuscular preparations, suppresses the
amplitude of excitatory synaptic potentials (EPSPs), and both depol-
arizes and reduces the input resistance of muscle fibers (Usherwood
and Machili, 1966; Yamamoto and Washino, 1980). The action of L-

glutamic acid is localized to circumscribed spots on the muscle
fiber, which are presumed to be junctional sites (Yamamoto and
Washio, 1980; Beranek and Miller, 1968). The reversal potentials of
the EPSP and the L-glutamic acid-induced depolarization are identical
(Yamamoto and Washio, 1980; Anwyl and Usherwood, 1974; Jan and Jan,
1976; Anwyl, 1977; Robinson, 1981, Irving and Miller, 1980; and
Irving and Miller, 1980).

The object of this study is to develop specific agonists for the
glutamate receptor on the insect muscle and to clarify the structural
constrations imposed on such agonists.

EXPERIMENTAL

The skeletal muscle fibers of the mealworm, Tenebrio molitor,
was used. Most of the experimental procedures described in this
paper have been published from our research group (Yamamoto and
Washio, 1980; Yamamoto and Washio, 1979; Fukami and Izawa, 1984; Usui
and Fukami, 1984; and Miyamoto, et al., 1985).

Typical changes of EPSP and membrane potentials at the mealworm
neuromuscular junctions during perfusion of glutamate agonists (L-
glutamic acid, quisqualic acid, L-asparatic acid) and some neuro-
active compounds (acetylcholone, curare, atropine, chlordimeform,
nereistoxin, fenitrothion and fenitroxon) are shown in Fig. 1.

The amplitude of EPSP decreased rapidly with marked depolarization of muscle membrane by applications of 5×10^{-5}M L-gutamic acid,
5×10^{-5}M quisqualic acid, and 2×10^{-3}M asparatic acid. These
reactions were restored almost completely after washing with the
drug-free standard saline and, therefore, reactions seem to be reversible (Fig. 1A,B,C).

The reason for the reduced amplitude of the EPSP is threefold.
First, the receptor for the natural transmitter was desensitized by
bath-applied glutamate agonists. Second, the driving force was
reduced owing to the depolarization induced by glutamate agonists.
Finally, the muscle input resistance was decreased by glutamate
agonists, resulting in a reduced IR.

With other compounds, responses were observed at higher concentrations. Acetylcholine did not suppress the amplitude of EPSP but
induced depolarization of muscle membrane at a concentration of 2×10^{-3}M. For further examination of this phenomenon, two cholinergic
blockers, atropine and curare, were used. Curare did not produce the
depolarization of muscle membrane but greatly suppressed EPSPs at a
concentration of 3.6×10^{-3}M, resulting from the block of the ion
channel linked to the receptor (Yamamoto and Washio, 1979). The
potentials were not restored completely after washing with drug-free

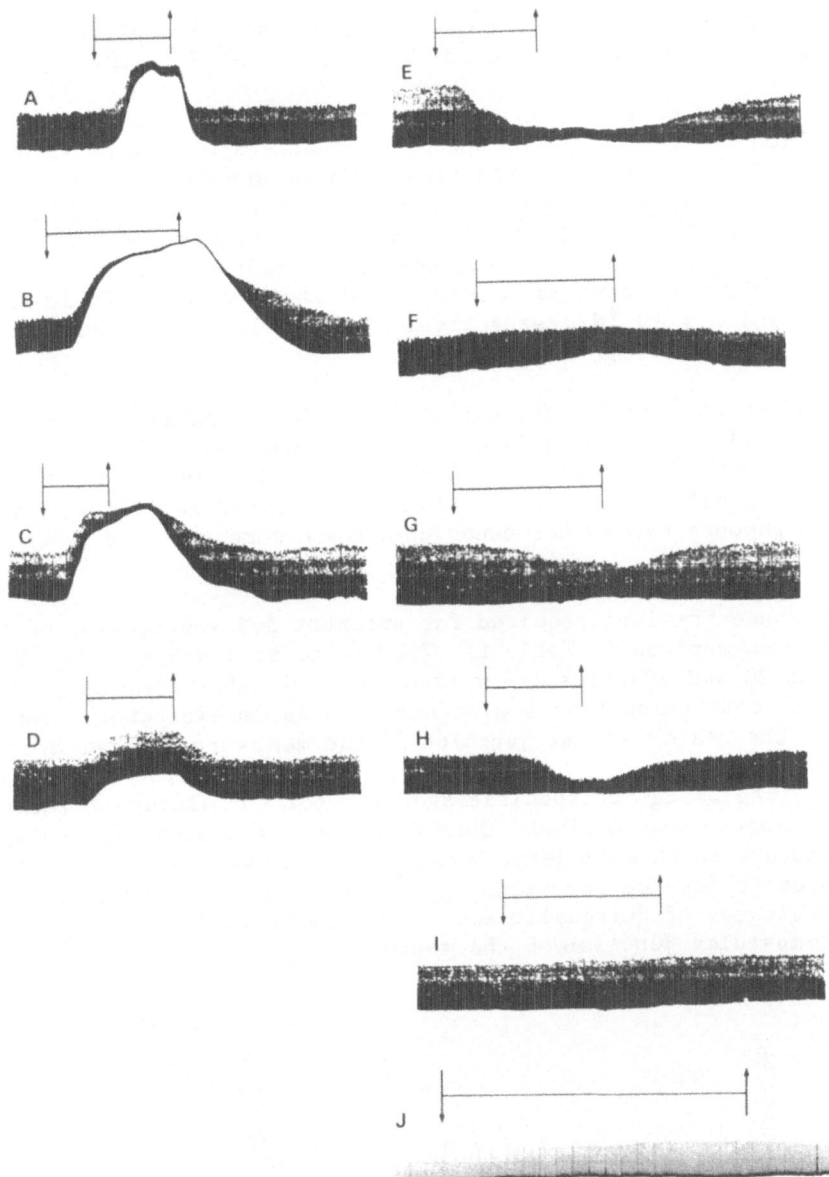

Fig. 1.   The action of some compounds on the membrane potentials and
EPSPs.   A:L-glutamic acid (5 x 10^{-5}M), B:quisqualic acid (5
x 10^{-5}M), C:L-asparatic acid (2 x 10^{-3}M), D:acetylcholine (2
x 10^{-3}M), E:curare (3.6 x 10^{-3}M), F:atropine (1 x 10^{-3}M),
G:chlordimeform (2 x 10^{-3}M), H:nereistoxin (2 x 10^{-3}M),
I:fenitrothion (5 x 10^{-4}M), J:fenitroxin (2 x 10^{-4}M).   The
downward arrow indicates the beginning of compound perfusion
and the upward arrow indicates the beginning of the washing.
Calibration, 10mV (A-H, J), 20mV (I), 1 min.

standard saline (Fig. 1E).   Atropine produced depolarization of
muscle membrane and suppression of EPSP at 10^{-3}M (Fig. 1F).

The effects of pesticides such as chlordimeform, nereistoxin,
fenitrothion and fenitroxon on the nerve-muscle system were also
examined.   Chlordimeform evoked both a slight depolarization of the
muscle membrane and suppression of EPSPs at 2×10^{-3}M (Fig. 1G), and
clear depolarization at 10^{-2}M.   Nereistoxin suppressed EPSP without
depolarization of the muscle membrane at a concentration of 2×10^{-3}M
(Fig. 1H).   Fenitrothion and fenitroxon at their maximum solubility
(5×10^{-4}M and 2×10^{-4}M, respectively) caused little increase in the
amplitude of EPSP (Fig. 1I,J).

Dose-response curves for L-glutamic acid, quisqualic acid and L-
asparatic acid are shown in Fig. 2.   Values are means of 3-6 determinations.   The concentrations inducing EPSP suppression were in the
order of quisqualic acid < L-glutamic acid < L-asparatic acid.   Similar dose-response curves for other compounds were obtained (data not
shown).

The concentrations required for apparent 50% suppression of EPSP
(IC_{50}) are summarized in Table 1.   Quisqualic acid and L-glutamic acid
were about 80 and 40 times lower than those of other compounds, supporting the contention that L-glutamic acid is an excitatory transmitter at the neuromuscular junction of the mealworm skeltal muscle.

Next, the design of specific agonists for the glutamate receptor
on insect muscle was studied.   Quisqualic acid has been used as a
lead structure in this project, because it is known to be the most
potent agonist for the junctional glutmate receptor.   One hundred
twenty analogues of quisqualic acid were synthesized and assayed at
the neuromuscular junction of the mealworm.

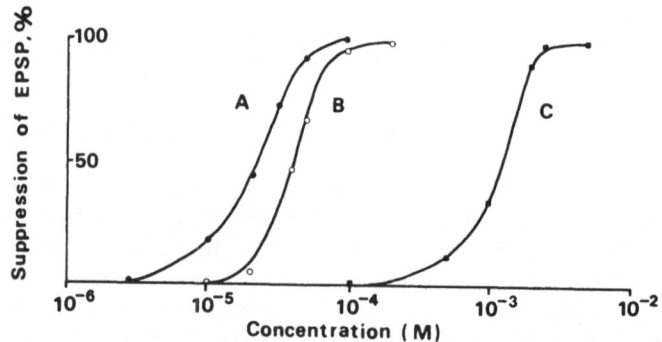

Fig. 2.   Dose-suppression curves of EPSP.   Each point represents
         the mean of 3-6 determinations. A:quisqualate,   B:L-
         glutamate, C:L-asparatate.

Table 1. The concentration for apparent 50% suppression of EPSP
amplitude (IC_{50}) and the effect on the depolarization of
muscle membrane

	IC_{50}	Depolarization
Quisqualate	2.3×10^{-5}	++
L-Glutamate	4.2×10^{-5}	++
L-Asparate	1.2×10^{-3}	++
Acetylcholine	-----	+
Atropine	1.8×10^{-3}	+
Curare	1.7×10^{-3}	-
Nereistoxin	1.6×10^{-3}	-
Chlordimeform	1.6×10^{-3}	+
Fenitrothion	-----	-
Fenitroxon	-----	-

In Fig. 3, the chemical structures of the compounds with agonistic action are shown. The compounds are arranged from top to bottom
in the order of their potency to induce depolarization of the muscle
membrane. Quisqualic acid was the most potent. L-Glutamic acid N-
thiocarboxyanhydride (L-GANTA) and DL-hydantoinpropionic acid (DL-
HPA) are new agonists for glutmate receptors discovered in our study.

The two agonists, L-GANTA and DL-HPA, had very similar effects
on this muscle preparation (Fig. 4A). They both produced a large
depolarization and suppressed the amplitude of EPSP. In Fig. 4A, the
time course of action of L-GANTA is shown on a slow time base. The
action of DL-HPA on the individual EPSPs is illustrated in Fig. 4B on
a fast time base. The depolarizing action of L-GANTA and DL-HPA
indicates that these two compounds activate subsynaptic amino acid
receptors which are linked to the ion channel.

The dose-depolarization curves for L-GANTA and DL-HPA are illustrated in Fig. 5. The apparent dissociation constant (K_d) was
approximately 7×10^{-4}M for L-GANTA and 9×10^{-4}M for DL-HPA. Comparing the results shown in Fig. 5 with those in Fig. 2, it can be
concluded that the dose-depolarization curves for L-GANTA and DL-HPA
lie between the glutamate curve and the asparatate curve.

The depolarization induced by L-GANTA or DL-HPA was accompanied
with a reduction in the input membrane resistance of the muscle
fiber. Fig. 6 shows a typical example of the current-voltage (I-V)
curves of a mealworm muscle fiber in the absence and presence of 5×10^{-4}M L-GANTA. This effect of L-GANTA on the I-V curve is quite
similar to that for L-glutamic acid (cf. Fig. 3 or 5). Essentially
the same result has been obtained for DL-HPA (data not shown).

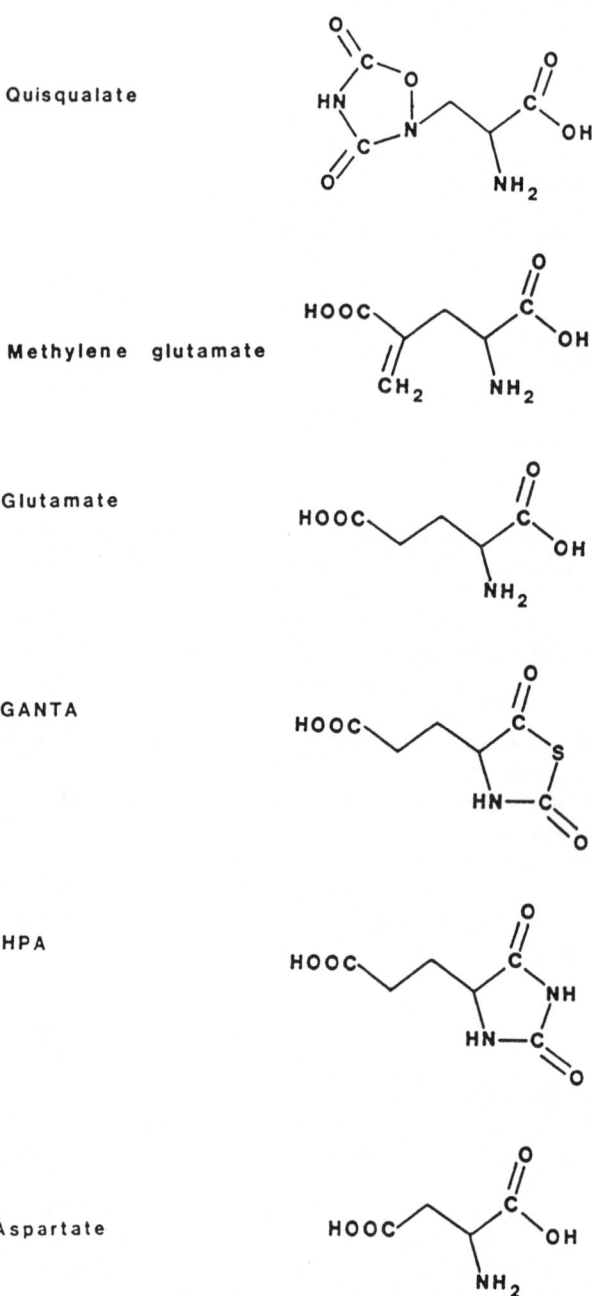

Fig. 3.   Chemical structure of glutamate agonists.   Compounds are
arranged from top to bottom in the order of potency.

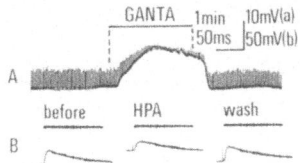

Fig. 4.   The action of two glutamate agonists. A: Depolarizing
          response of L-GANTA is shown. Upward deflections are EPSPs
          evoked at 0.25 Hz and base line means resting membrane
          potential. B: Depolarizing response of DL-HPA. EPSPs
          before, during and after the application are shown. Upper
          trace means 0 potential. Note the similar action of L-GANTA
          and DL-HPA.

     We have examined some hydantoin analogues related to GANTA and
HPA including methyl-DL-hydantoin propionate, methyl-DL-2-thionhydan-
toin propionate, 1,5-tetramethylene hydantoin-1'-carboxylic acid, 3-
n-butyl-1,5-tetramethylene-hydantoin-1'-carboxylic acid, and 1,5-
tetramethylene hydantoin-4'-carboxylic acid. None of these could
induce depolarization accompanied by suppression of EPSPs in the
muscle. Quisqualate derivatives with modified ω-acid group were

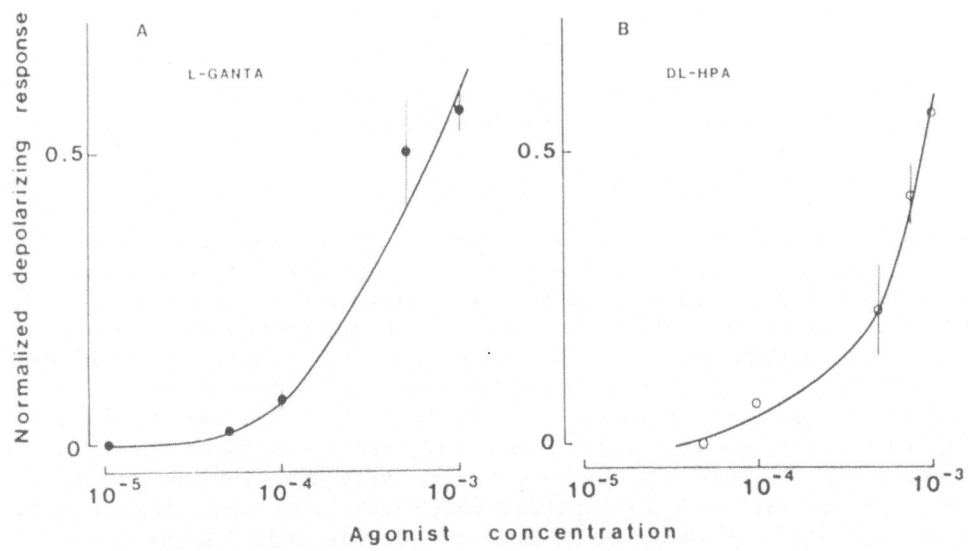

Fig. 5.   Dose-depolarization curves for L-GANTA (A) and DL-HPA (B).
          The normalized response is defined as a depolarization
          magnitude (mV) divided by driving force (mV). The driving
          force is the difference between the resting membrane
          potential before the application and the reversal potential
          (-10mV) for glutmate.

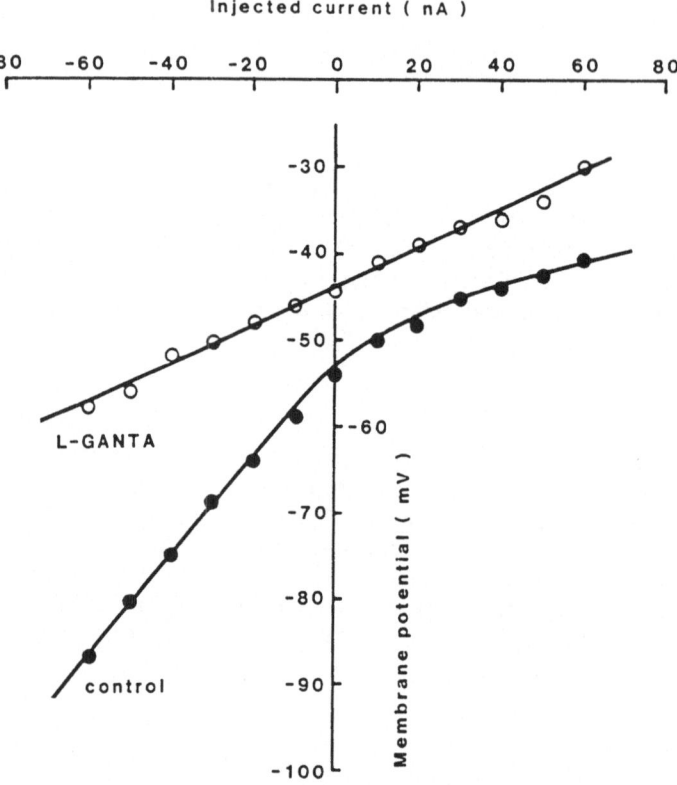

Fig. 6.   Current-voltage curves in the presence and absence of 5 x
          10^{-4}M L-GANTA.   Note a reduction in the input resistance by
          L-GANTA.

generally inactive on the mealworm neuromuscular junction.   For example, compound 1-5 (Table 2), where one of the carbonyl groups was
esterified or replaced with some other functional group, had no
effect either on the EPSP or on the resting membrane potential.
Quisqualate derivatives such as compound 6 to 11 (Table 2), obtained
by replacing the oxygen with a sulfur atom in the hetrocyclic ring,
lacked any agonistic activity.   Compounds 12 to 18 in Table 2 were
ω-amide derivatives of glutamic acid and were found to be inert in our
preparation.   In addition, we have tested many tyrosine and trypto-
phan derivatives, and oligopeptides which have structures common with
quisqualic acid, glutamic acid, and/or ibotenic acid.   These compounds did not show any agonistic activity.

DISCUSSION

    L-Glutamic acid, quisqualic acid and L-asparatic acid produced
depolarization of the muscle membrane at the neuromuscular junction

Table 2. Chemical structures of quisqualate and glutamate analogues

$$X_3 - \overset{\underset{\displaystyle X_4}{\big|}}{\underset{X_2}{C}} - X_1 - \overset{\underset{\displaystyle NH_2Z}{\big|}}{CH} - CO_2Y$$

Compounds	X_1	X_2	X_3	X_4	Y	Z
1	O	O	CH_2	$-CO_2C_2H_5$	H	-
2	O	O	NH	$-CO_2C_2H_5$	H	-
3	O	O	NH	$-CO - \phi^1$ (F, F)	H	-
4	O	O	NH	$-CH_3$	CH_3	HBr
5	O	O	NH	$-CO- \phi$ (F, F)	CH_3	HBr
6	O	S	NH	$-CO- \phi$	CH_3	HBr
7	O	S	NH	$-COCH_3$	CH_3	HBr
8	S	O	NH	$-CO_2C_2H_5$	CH_3	HBr
9	S	O	NH	$-CO- \phi$	CH_3	HBr
10	S	O	NH	$-CH_3$	CH_3	HBr
11	S	O	NH	$-CO- \phi$ (F, F)	CH_3	HBr
12	CH_2	O	NH	$-CH_3$	H	HCl
13	CH_2	O	NH	$i-C_3H_7$	H	HCl
14	CH_2	O	NH	$-\phi$	H	HCl
15	CH_2	O	NH	$\phi -Cl$	H	HCl
16	CH_2	O	NH	$\phi \underset{Cl}{\overset{Cl}{<}} Cl$	H	-
17	CH_2	O	NH	$\phi -COOH$	H	-
18	CH_2	O	NH	$-NHCO_2C_2H_5$	H	HCl

[1] ϕ represents a benzene ring.

of the mealworm in the same manner, indicating that they act as agonists for the amino acid receptor. On the other hand, acetylcholine did not suppress EPSP, but did produce depolarization. Is acetylcholine acting on the neuromuscular junction of the mealworm? Such a possibility is extremely unlikely because of the following reasons. First, the concentration of acetylcholine was very high (2 x 10^{-3}M). Second, cholinergic blockers such as curare did not show the blocking action for the natural receptor, but for the ion channel which is linked to the receptor (Yamamoto and Washio, 1979). Third, the inhibitors for acetylcholine esterase such as fenitrothion were less effective. Considering the results, it is suggested that L-

glutamic acid is the most probable candidate for the natural excitatory transmitter at the neuromuscular junction of the mealworm.

It has been reported that L-glutamic acid acts on the locust
femur muscle, inducing contraction at the lowest concentration among
various amino acids and their related compounds (Usherwood and
Machili, 1968). The IC_{50} values of L-glutamic acid and L-asparatic
acid for the neurally evoked contraction of locust femur muscle were
6.4×10^{-5}M and 1.5×10^{-5}M, respectively (Clement and May, 1974).
These concentrations were similar to those obtained in our experiments, 4.2×10^{-5}M for L-glutamic acid and 1.2×10^{-3}M for L-
asparatic acid. In comparing L-glutamic acid with other glutamate
agonists on the neuromuscular transmission in insects and crustacea,
relative potencies of these agonists were in the order of quisqualic
acid > L-glutamic acid > asparatic acid (Clement and May, 1974;
Shinozaki and Shibuya, 1974; Clark et al., 1979; and Gration et al.,
1981).

It has been shown in the current study that new quisqualate
analogues, L-GANTA and L-HPA, have an agonistic activity at the
postsynaptic amino acid receptor on insect skeletal muscles. Many
other quisqualate analogues, however, had little or no effect on the
neuromuscular junction of the mealworm.

The minimum structural requirement for a molecule to have an
agonistic activity at the mealworm neuromuscular junction is that the
compound has an amide or amino side chain at one end, separated from
a carboxyl group at the ω-position by two intervening carbons or
nitrogens. Quisqualic acid lacks an ω-carboxyl, but the carbonyl
groups surrounding the nitrogen may allow the nitrogen to ionize.
This suggests that the agonist molecule must have a negative charge
at each end of the main chain and positive charge at the N position.
These features of agonists for the amino acid receptors in the mealworm neuromuscular junction are strikingly similar to those found in
the locust muscle (Usherwood and Machili, 1966; Clements and May,
1974) and in vertebrate central neurons (Watkins and Evans, 1981).

Strickly speaking, amino acid receptors in the nervous system
may be further divided into several subgroups with respect to their
relative affinity to agonists (Curtis and Watkins, 1960; Takeuchi and
Takeuchi, 1964; and Watkins, 1980). It would be of great interest to
see how L-GANTA and DL-HPA interact with each of these receptor
types.

REFERENCES

Anwyl, R., 1977. Permeability of the post-synaptic membrane of an
    excitatory glutamate synapse to sodium and potassuim, J.
    Physiol., 273:367.

Anwyl, R., and Usherwood, P.N.R., 1974. Voltage clamp study of
    glutamate synapse, Nature, 252:591.
Beranek, R., and Miller, P.L., 1968, The action of iontophoretically
    applied glutamate on insect muscle fibres, J. Exp. Biol., 49:83.
Clark, R.B., Bration, K.A.F., and Usherwood, P.N.R., 1979, Agonist-
    dependent desensitization of glutmate receptor at locust nerve-
    muscle junction, J. Physiol., 295:93.
Clement, A.N., and May, T.E., 1974, Pharmacological studies on a
    locust neuromuscular junction, J. Exp. Biol., 61:421.
Clements, A.N. and May, T.E., 1974, Studies on locust neuromuscular
    physiology in relation to glutamic acid, J. Exp. Biol., 60:673.
Curtis, D.R. and Watkins, J.C., 1960, The excitation and depression
    of spinal neurones by structurally related amino acid, J.
    Neurochem., 6:117.
Fukami, J., and Izawa, N., 1984, Mode of action of glutamate
    agonists and related insecticides on the muscle contraction and
    nerve-muscle junction of insects, in: "Cellular and Molecular
    Neurotoxicology," T. Narahashi, ed., Raven Press, New York.
Gration, K.A.F., Lambert, J.J., Ramsey, R.L., Rand, R.P., and
    Usherwood, P.N.R., 1981, Agonist potency determination by patch
    clamp analysis of single glutamate receptors, Brain Res.,
    230:400.
Irving, S.N., and Miller, T.A., 1980, Asparatate and glutamate as
    possible transmitters at the "slow" and "fast" neuromuscular
    junctions of the body wall muscles of Musca larvae, J. Comp.
    Physiol., 135:299.
Jan, L.Y., and Jan, Y.N., 1976. L-Glutamate as an excitatory tran-
    smitter at the Drosophila larval neuromuscular junction, J.
    Physiol., 262:215.
Miyamoto, T., Oda, M., Yamamoto, D., Kaneko, J., Usui, T., and
    Fikami, J., 1985, Agonistic action of synthetic analogues of
    quisqualic acid at the insect neuromuscular junction, Arch.
    Insect Biochem. Physiol., in press.
Nistri, A., and Constanti, A., 1979, Pharmacological characteriza-
    tion of different types of GABA and glutamate receptors in
    vertebrates and invertebrates, Prog. Neurobiol., 13:117.
Robinson, N.L., 1981, Glutamate as the transmitter at fast and slow
    neuromuscular junctions of larval Diptera, J. Comp. Physiol.,
    144:139.
Shinozaki, H., and Shibuya, I., 1974, A new potent excitant, quis-
    qualic acid: Effect in crayfish neuromuscular junction, Neuro-
    pharmacol., 13:665.
Takeuchi, A., and Takeuchi, N., 1964, The effect on crayfish muscle
    of iontophoretically applied glutamate, J. Physiol., 170:296.
Usherwood, P.N.R., 1978, Amino acids as neurotransmitters, Adv.
    Comp. Physiol. Biochem., 7:227.
Usherwood, P.N.R., and Machili, P., 1968, Chemical transmission
    at the insect excitatory neuromuscular synapse, Nature,
    210:634.

Usherwood, P.N.R., and Machili, P., 1968, Pharmacological properties of excitatory neuromuscular synapses in the locust, J. exp. Biol., 49:341.

Usui, T., and Fukami, J., 1984, The effects of L-glutamic acid, glutamic agonists, acetylcholine and several drugs on the excitatory post-synaptic potentials at the neuromuscular junction of the larval mealworm, Tenebrio molitor Linne (Coleoptera; Tenebrioidae), Appl. Ent. Zool., 19:151.

Watkins, J.C., 1980, New light on amino acid-mediated synaptic excitation, TINS, 3:61.

Watkins, J.C., and Evans, R.H., 1981, Excitatory anion acid transmitters, Ann. Rev. Pharmacol. Toxicol., 21:165.

Yamamoto, D., and Washio, H., 1979, Curare has a voltge dependent blocking action on the glutamate synapse, Nature, 281:372.

Yamamoto, D., Washio, H., and Fukami, J., 1983, Evidence for a presynaptic action of chlordimeform at the insect neuromuscular junction, Arch. Insect Biochem. Physiol., 1:33.

Yamamoto, D., and Washio, H., 1980, L-Glutamate as an excitatory transmitter at the neuromoscular junction of a beetle larvae, J. Insect Physiol., 26:253.

INDEX